A Naturalist's Guide to Plant Communities of Pacific Northwest Dune Forests and Wetlands

Many secluded areas in Pacific Northwest dune forests are perfect for study and contemplation.

A Naturalist's Guide to Plant Communities of Pacific Northwest Dune Forests and Wetlands

GEORGE POINAR, JR.

BRIT
PRESS
2019

A Naturalist's Guide to Plant Communities of Pacific Northwest Dune Forests and Wetlands

First edition 2019
Printed in Korea; prepress U.S.A.

ISBN: 13-978-1889878-54-6

Production: Rebecca Horn, rhorngraphics, rlrchorn@verizon.net
Cover Design: Rebecca Horn, rhorngraphics

Front Cover:
Top: Dune forest surrounding a dune lake in Oregon.
Bottom *(left to right from top to bottom)*: Great blue heron, salmonberry flowers (top), Timarcha leaf beetle (bottom), Sitka spruce bough, morel fungus, black bear, spittlebug nymph. Photos by George Poinar, Jr. except black bear taken by the late Henry Heil.
Back cover (left to right from top to bottom):
Drumming katydid, narrowleaf bur-reed flower, yellow-headed sawfly larva, red-winged blackbird, red-headed soldier beetle, white water lily.
Photos by George Poinar, Jr.

How to cite:
Poinar, G., Jr. 2019. A naturalist's guide to plant communities of Pacific Northwest dune forests and wetlands. Sida, Bot. Misc. 50. Botanical Research Institute of Texas, Fort Worth, Texas, U.S.A.

A Naturalist's Guide to Plant Communities of Pacific Northwest Dune Forests and Wetlands is published with support of the author and the Botanical Research Institute of Texas Press.

Distribution of copies by:
Botanical Research Institute of Texas Press
1700 University Dr.
Fort Worth, Texas 76107-3400, U.S.A.
Telephone: 1.817.332.4441; Fax: 1.817.332.4112
Website: http://shop.brit.org
Email: orders@brit.org

BRIT
PRESS
Botanical Miscellany, No. 50
ISSN: 0833-1475
1 April 2019

A Naturalist's Guide to Plant Communities of Pacific Northwest Dune Forests and Wetlands

A BOOK TO PROMOTE BOTANICAL AND ENVIRONMENTAL EDUCATION, SCIENTIFIC EXPLORATION, AND CONSERVATION IN THE PACIFIC NORTHWEST

DEDICATION

I dedicate this book to my loving wife, Roberta,
for her devotion, support,
and encouragement over the years.

An enchanted scene in a Pacific Northwest dune forest.

TABLE OF CONTENTS

PREFACE

While studying strand and sand dune plant communities along the Pacific Northwest coast (Poinar 2016), I discovered that many of the sandy areas extended inland for some distance. In some cases, these inland "dune" areas were covered with large evergreen forests and surrounded by wetlands. Examination of plant communities associated with these dune forests and wetlands showed they were unique in many ways, and contained vegetation not found in other North American woodlands.

In undisturbed areas along the Pacific Northwest coast, a walk inshore from the ocean at low tide traverses the strand and leads to the sandy foredunes. Here are found plants that face the rigorous ocean climate throughout the year. Continuing on from the foredunes reaches the back sand dunes that are dominated by grasses, shrubs, and small trees. Still farther inland is where dune forests with their associated ponds, lakes, bogs, seeps, marshes, and swamps are found. These various wetlands, with their own plant communities, can be quite ephemeral or last for decades. How much longer these dune forests and wetlands will exist depends to a great extent on our ability to preserve these regions.

The present work examines plant communities of Pacific Northwest dune forests and wetlands along with their invertebrate and vertebrate associates. While not all of the plants communities could be included, those examined shows how intricate and fascinating life can be in a relatively unexplored habitat so close to civilization.

Having a laboratory at Waldport, Oregon, provided the author with an opportunity to study plant communities in dune forests and wetlands from Washington to northern California. Included here are not only plants native to the Pacific Northwest but also species from outside the area that have become naturalized. Hopefully, exploring these forests and associated wetlands will be as enjoyable to the reader as it was to the author.

ACKNOWLEDGMENTS

This book would never have been written without the love, support, and understanding of my dear wife, Roberta. Together, we made countless excursions along the Pacific coast and spent numerous hours discussing the various types of life found in dune forest and wetland habitats.

I would like to thank the following persons for their assistance in identifying various biota: Don Bright, John W. Brown, Daniel Burckhardt, Kenton Chambers, Donald Davis, Julian P. Donahue, John Doyen, Raymond Gagne, Dick Goeden, Pete Haggard, Richard Halse, Paul Hammond, John Heppner, Ron Hodges, Thomas Horton, Tom Kaye, Jerry Krantz, James LaBonte, the late Jack Lattin, Andrei Legalov, Chuck Meissner, Jeff Miller, Andy Moldenke, Alfred Newton, Allen Norrbom, Barry O'Connor, Gary Parsons, Jerry Powell, the late Frank Radovsky, D. Christopher Rogers, Dana Ross, Lynn Royce, David Smith, Jeffrey Stone, Scott Shaw, Margaret Thayer, Don Thomas, David J. Voegtlin, Barbara Wilson, Richard Worth, and Richard Wescott.

I am also grateful to the following persons who introduced me to specific dune forest localities or provided other courtesies: Ralph Berry, Alex Brown, Pete Haggard, Andrea Pickart, David Schlesinger, the late Alfred Wiedemann, and my children and grandchildren. All photos were taken by the author with the exception of the black bear that was photographed by the late Henry Heil.

Lofty conifers in primeval Pacific Northwest dune forests were once commonplace.

INTRODUCTION TO DUNE FORESTS

Walking inland from the sea beyond the sand dunes along the Pacific Northwest coast brings you to the inviting dune forests. Characteristics of the soil and vegetation of these forests differ from those of other woodlands. While most dune forests occur near the coast, some extend a fair distance from the sea. The sandy soil that is characteristic of these forests originally was an extension of the coastal sand dunes. Over time, various types of vegetation adapted to the climatic and geologic changes that occurred during the Pleistocene epoch.

Understories of shrubs and herbs in Pacific Northwest dune forests have unique associations not found in other North American woodlands.

Dune forests do not have sharply contrasted seasons as in broadleaf forests where falling leaves and freezing temperatures are common during part of the year. Aside from the flowers that appear in the spring and summer, the predominantly evergreen vegetation is fairly uniform for the remainder of the year. The plants are not affected by ocean winds and salt spray, and the periodically occurring fog banks maintain a cool, moist environment.

Hikers often must use narrow trails to traverse through the dense vegetation of dune forests.

A variety of unique plant communities and their associated herbivores occur in dune forests. Here are trees supporting succulent lichens, fallen logs covered with mosses and liverworts, and tangled bushes lashed together with impassable vines. Captivating century-old conifers like Sitka spruce, Douglas-fir,

Towering conifers dominate the landscape in many climax dune forests.

Assemblages of conifers in dune forests often border lakes and ponds.

and western hemlock fill the air with resinous scents and are at home in the cool, wet climes of these Pacific Northwest dune forests. These evergreens often dominate the landscape and their colossal trunks can tower up for more than a hundred feet, straight as an arrow. While broadleaf trees also flourish within the confines of these forests, they are rarely dominant.

The stunning beauty of Pacific Northwest dune forests, especially those bordering lakes, makes wandering through them a memorable event. Nature lovers will find a striking display of fascinating plants and animals. Unfortunately, the habitats of these forests are being upset by lumbering, construction, and the establishment of non-native plants. Human activity is the most serious threat to dune forests today.

In the recent past, Pacific Northwest conifers supplied a number of survival items for Ice Age gatherers. Wood and bark were used as fuel as well as for making spoons, bowls, spear handles, harpoon shafts, and fishhooks. Seeds, shoots, and inner bark supplied food, and teas made from bark were taken for internal injuries. Branches were used to construct small dwellings and other tree products were used for tanning leather and making baskets and fishing nets. Pitch was used for treating colds, caulking canoes, waterproofing containers, and making torches.

All trees have insect pests and conifers are no exception. While confers may have evolved resin production to repel pests, many insects have adapted to this feature. Thus, it is possible to find insects nestled among the needles, hidden under the bark or buried within the wood of these trees. Many of these and other insects associated with dune forest vegetation, as well as selected invertebrates and vertebrates, are included in the present work. The following values are used to indicate the size of various inhabitants of these forests: microscopic = under 1 mm; small = 1–10 mm; midsized = 10–20 mm; large = 20–30 mm and very large = over 30 mm.

DUNE FOREST PLANT COMMUNITIES

RED ALDER

Red alder can be found growing among conifers in dune forests, especially along the border of wetlands. It is one of the few hardwood or broadleaf trees found in dune forests. Red alder is a fast grower, possibly out of necessity, since it must maintain itself in spite of many herbivores. The red dye and antibiotic properties in its bark don't seem to deter many insects, but they aren't the only creatures that feed on these trees. I remember returning to a site where I had examined a red alder tree the previous day, only to find it gone. All that was left was a severed trunk and a few broken twigs. The beavers had been quite busy and since this was one of the last red alders in the vicinity of their pond, I wasn't surprised when the beavers were gone the following year.

One curious feature of red alder is that the edges of the leaves are bent over and attached to the under portion of the leaf. The bent portion might as well be glued since it is impossible to straighten without tearing the leaf. The adaptive value of this character is unknown, although it may allow rain droplets to run quickly off the leaves.

What is amazing about insects that occur on red alder is their range in color. Bluish-purple flea beetles that gather in small groups to sun themselves on the leaf tips may be adjacent to black-spotted orange leaf beetles. Some alder caterpillars are green while others are adorned with white lines or spots and still others are mottled tan to dark brown, some even with a pair of white "eye spots," the purpose of which is a mystery. Such a variety of colors against the green alder leaves provide a visual dissonance. Since it seems that these colors and markings would make them easy prey for birds and other predators, why did they evolve? Many caterpillars restrict their rapid feeding activities to the night hours so that in the morning only damaged leaves remain, with no sign of the perpetrators. Many caterpillars that feed continuously on a single leaf construct their own lodgings by twisting the tips of the leaves back over themselves, thus creating a "leaf roll." These are just some of the interesting aspects of the red alder community.

RED ALDER COMMUNITY Herbivores	Detritivores/Predators	Parasite
Leaf roller	Wasp mimic crane fly	Leaf blister fungus
Crossed wing stink bug	Awl fly	
White spotted sawfly	Wood louse	
Tufted thyatirid moth		
Mottled caterpillar		
Banded alder borer		
Flea beetle		
Leaf beetle		
Case bearer		

RED ALDER

(*Alnus rubra:* Betulaceae)

Characteristics: Fast growing, short-lived, deciduous; light gray bark. Leaves elliptical with coarse teeth. Flowers appear as hanging catkins before leaves have fully developed. Fruit small, oval cones that overwinter on tree.

Regional habitat: Wet areas of dune forests along the Pacific coast from Washington to central California.

Comments: Bacteria in root nodules transfer atmospheric nitrogen into forms used by the tree. Red alder is the principal tree used by beavers for food and shelter.

LEAF ROLLER

(*Epinotia solandriana*: Lepidoptera: Tortricidae)

Characteristics: Caterpillar pale yellow to greenish-gray; dark head capsule. Adult mainly brownish with diamond-like pattern on back when wings are closed.

Regional habitat: Dune forests from Washington to California.

Comments: These caterpillars also attack willows. Wing colors and patterns are quite variable.

WASP MIMIC CRANE FLY

(*Ctenophora vittata angustipennis*: Diptera: Tipulidae)

Characteristics: Larva gray to pale white. Adult large, shiny; thorax black and yellow; abdomen long, black and red.

Regional habitat: Widespread throughout the Pacific Northwest.

Comments: Larvae feed on decaying wood of alder and other deciduous trees. Wasp-like features of adults may protect them from predators.

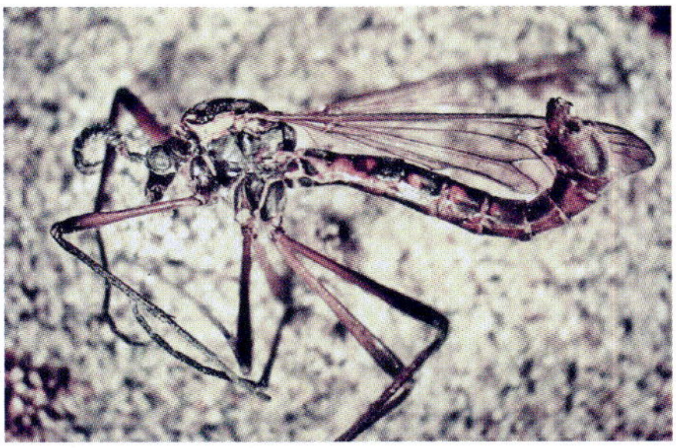

CROSSED WING STINK BUG

(*Elasmostethus cruciatus*: Hemiptera: Acanthosomatidae)

Characteristics: Juvenile colorful, almost round. Adult large, oval; upper parts greenish, under parts yellowish; antennae long; wings with cross pattern.

Regional habitat: Throughout the Pacific Northwest.

Comments: The wide cross on its back gives the bug its common name. Adults are frequently found on alder cones. Eggs and early stage juveniles are on leaves. Newly hatched juveniles often remain together after hatching (homotypic agglutination).

WHITE SPOTTED SAWFLY

(*Trichiosoma triangulum*: Hymenoptera: Cimbicidae)

Characteristics: Caterpillar light green with transverse rows of white spots; head whitish with reddish-brown areas. Adult black; eyes large; antennae with knobs on tips; wings brownish-tinted, shiny.

Regional habitat: Caterpillars on leaves of alders and willows in dune forests from Washington to California.

Comments: Color patterns on the caterpillar head vary considerably. Sometimes the head is completely white.

TUFTED THYATIRID MOTH
(*Pseudothyatira cymatophoroides*: Lepidoptera: Drepanidae)

Characteristics: Caterpillar reddish-brown, thin black line running length of back, usually pair of white spots on side of body near head. Pupa bulky, dark green. Adult brown; forewings with dark spots.

Regional habitat: Forests along the Pacific coast from British Columbia to California.

Comments: The function of the white spots on the sides of these caterpillars is unknown.

LEAF BLISTER FUNGUS
(*Taphrina* sp.: Ascomycota: Taphrinaceae)

Characteristics: Affected red alder leaves become puffy and swollen with reddish pustules or blisters. Spores form on surface of hypertrophied leaves.

Regional habitat: Dune forests from Washington to California.

Comments: Swollen leaf areas are filled with fungal hyphae.

WOOD LOUSE

(*Ligidium gracile*: Isopoda: Ligiidae)

Characteristics: Shiny, reddish brown, elliptical; seven pairs of legs; terminal appendages (uropods) paired; antennae apical with 10 segments.

Regional habitat: Moist areas with an abundance of leaf litter along the coast from Washington to California.

Comments: These scavengers feed on damp decaying wood or other plant matter, but seem to be especially fond of alder wood. Wood lice are mostly active at night and remain concealed in decaying leaf litter during the day.

MOTTLED CATERPILLAR

(*Lithophane* sp.: Lepidoptera: Noctuidae)

Characteristics: Caterpillar large, mottled brownish, dark brown V-shaped pattern extending down back; head tan. Adult moths were not observed.

Regional habitat: Washington to California.

Comments: Other noctuid caterpillars that feed on red alder are the green *Orthosia hibisci*, the white-lined green *Orthosia pacifica*, and the cream-colored *Mesogona* sp.

BANDED ALDER BORER

(*Rosalia funebris*: Coleoptera: Cerambycidae)

Characteristics: Adult large, slender with long legs; thorax white with black diamond in middle; antennae as long as body; wing covers (elytra) black with two or three transverse white bands.

Regional habitat: Dune forests from Washington to California.

Comments: Legless, white larvae bore into red alder wood. Adult beetles can be found on flowers or resting on fallen logs. Nematodes and mites are carried under the wing covers.

FLEA BEETLE

(*Macrohaltica ambiens*: Coleoptera: Chrysomelidae)

Characteristics: Egg yellowish, elongate with rounded ends. Larva black, 2nd and 3rd thoracic segments with yellow areas; legs short. Adult elongate, shiny, bluish-purple to black; antennae fairly long.

Regional habitat: On red alder in dune forests from Washington to California.

Comments: Adults overwinter in debris and emerge in spring to feed, mate, and lay egg clusters on leaves. Adults jump rapidly, often ending the leap with a flight. They frequently mass together in mating pairs on leaves during the day.

LEAF BEETLE

(*Chrysomela mainensis*: Coleoptera: Chrysomelidae)

Characteristics: Egg pale yellow, elliptical. Larva black, elongate. Pupa yellow, shiny. Adult ovate; head black, antennae short, reddish; thorax with central black area; wing covers reddish-orange with black spots (similar to those on some ladybird beetles).

Regional habitat: Dune forests throughout the Pacific Northwest.

Comments: Adults overwinter in debris and emerge in spring to feed, mate, and oviposit on leaves.

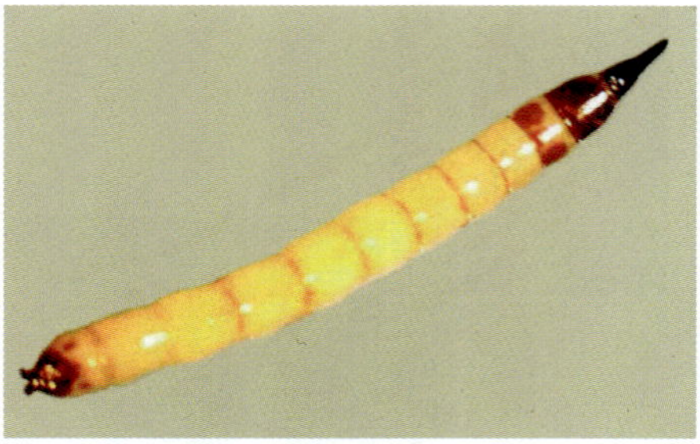

AWL FLY

(*Xylophagus* sp.: Diptera: Xylophagidae)

Characteristics: Larva elongate, yellowish; head pointed, black; thorax and posterior plate dark with two protruding spiracles; legless. Adult elongate, narrow; legs long; head small with proportionally large eyes.

Regional habitat: Dune forests throughout the Pacific Northwest.

Comments: Larvae, which are distinctive with their pointed black heads and yellow bodies, can be found in rotting wood of red alder, especially in trees felled by beavers. They are predators, feeding on other insects in the wood.

CASE BEARER

(*Coleophora pruniella*: Lepidoptera: Coleophoridae)

Characteristics: Caterpillar small, grayish; found inside tan portable case made from leaf litter. Adult yellowish; antennae long; wings with long narrow scales extending from rear margins.

Regional habitat: Washington to California.

Comments: This case bearer also occurs on willows, poplars, and stone fruits. A careful examination of the caterpillar case will often reveal venation patterns of the host plant.

HAIRY MANZANITA

Tearing the tough, leathery leaves of hairy manzanita is not easy, and that may be why the foliar insects I found on this plant don't attempt to eat the entire leaf but just nibble away on the top layer. Some insects protect themselves from predators by folding over just enough of the leaf to cover themselves. This technique is used by leaf fold and lined caterpillars on hairy manzanita.

The delicate little berry weevil centers its life cycle on the fruit. Larvae develop within the berry and form their pupae there. Newly formed adults leave through a hole they make in the tip of the fruit.

Stink bugs are well known for their ability to emit foul odors that ward off enemies. They can be herbivores or carnivores and one of each occurs on hairy manzanita. Stink bugs have a triangular shape with a broad thorax extending back to the narrow tip of the abdomen. While the herbivorous brown stink bug is uniformly light tan and blends in against dried leaves, the predatory red-banded stink bug displays red lines on its back. Such bold markings may warn potential enemies that it has a foul taste.

HAIRY MANZANITA COMMUNITY Herbivores	Predator
Brown stink bug	Red-banded stink bug
Berry weevil	
Leaf fold moth	
Lined caterpillar	

HAIRY MANZANITA
(*Arctostaphylos columbiana*: Ericaceae)
Characteristics: Tall, bushy, evergreen shrub. Leaves hairy, alternate, ovate. Flowers white to pink, urn-shaped (like blueberry flowers). Fruits reddish-brown berries.
Regional habitat: Forest clearings along the Pacific Northwest coast from Washington to California.
Comments: Berries are edible but may cause constipation.

BROWN STINK BUG

(*Cosmopepla intergressus*: Hemiptera: Pentatomidae)

Characteristics: Adult oval, light brown; eyes small; antennae slender.

Regional habitat: Along the Pacific coast from Washington to California.

Comments: This stink bug feeds on berries of hairy manzanita.

BERRY WEEVIL

(*Anthonomus* sp.: Coleoptera: Curculionidae)

Characteristics: Larva small, white to yellow; head tan; legless. Pre-pupa similar in color. Adult brownish-black; snout fairly long.

Regional habitat: Forest edges from Washington to California.

Comments: Larvae develop and pupate within the berries. Adults emerge from the berries in late summer (note exit hole in berry) and overwinter in the debris around the base of the plant.

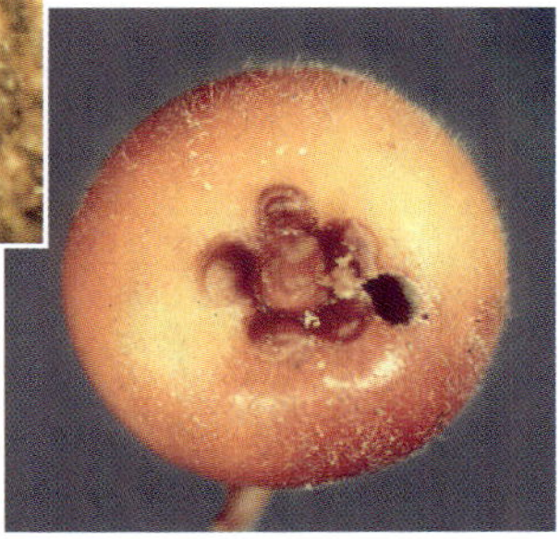

LEAF FOLD MOTH

(*Ethmia* sp.: Lepidoptera: Ethmiidae)

Characteristics: Adult small, grayish-brown with white "shoulder" spots; wings narrow with pale gray band running along back (noticeable when at rest).

Regional habitat: Dune forests in California and Oregon.

Comments: Caterpillars live within leaves they web together.

LINED CATERPILLAR

(*Mesogona* sp.: Lepidoptera: Noctuidae)

Characteristics: Young caterpillar cream; brown dorsal line running length of abdomen; head capsule tan. Mature caterpillar pink to dark brown; head brown; dark patch behind head. Adult uniformly tan, gray, or reddish-brown with few markings; hind wings sometimes pinkish.

Regional habitat: Dune forests in Oregon and California.

Comments: Caterpillars develop inside folded leaves but still fall prey to the red-banded stink bug.

RED-BANDED STINK BUG

(*Perillus splendidus*: Hemiptera: Pentatomidae)

Characteristics: Adult short, stout. black; pronotum with red bands often forming geometric patterns; head small; antennae short.

Regional habitat: Dune forests in Oregon and California.

Comments: This stink bug is predaceous and attacks other insects besides the lined caterpillar.

BEARBERRY

If you come across a low shrub in some forest clearings with evergreen, leathery leaves, white urn-shaped flowers, and red berries, it is most likely bearberry. Although the fruit is rather tasteless to us, bears avidly eat these ripe summer berries. Native Americans are said to have dried and then smoked the leaves, perhaps for narcotic or medicinal effects.

Small gall-producing aphids are common on bearberry leaves, turning leaf tissue into a cancerous mass of cells. What chemical substance the female aphid injects into the plant tissues to initiate such a response is unknown. Slowly, plant cells enlarge and grow completely around the feeding aphids, thus sealing them off from the outside. These red galls are quite conspicuous and initially appear to be colorful flowers or fruit. Insect galls usually contain a single individual, but bearberry galls contain an entire family of sometimes over 10 individuals, with hardly enough room to turn around. One unknown is how the aphids get rid of their waste, which might be absorbed and even utilized by the gall tissue itself. Opening one of these galls, which can only be done by tearing or cutting away part of the wall, exposes one or two aphid females with their offspring. In late summer, winged adults also are present, but how they exit the gall is unknown.

BEARBERRY COMMUNITY Herbivores	Symbiont	Commensal
Leaf tip moth	Bolete	Reindeer lichen
Cutworm		
Leaf gall aphid		

BEARBERRY
(*Arctostaphylos uva-ursi*: Ericaceae)
Characteristics: Evergreen, low growing shrub forming long rooting branches. Leaves alternate, oval, shiny, leathery with smooth margins. Flowers pinkish-white, urn-shaped; borne in small clusters. Fruits red berries with large seeds.

Regional habitat: Openings by the edges of dune forests along the Pacific Northwest coast.

Comments: Native peoples dried the leaves for smoking and medicine. Bears eat the berries, giving the plant its common name.

LEAF TIP MOTH

(*Epinotia* sp.: Lepidoptera: Tortricidae)

Characteristics: Caterpillar white to yellow with some erect body hairs. Adult slender; wings with alternating brown and whitish areas (noticeable when resting).

Regional habitat: Forest clearings from Washington to California.

Comments: Caterpillars live inside folded leaves positioned at the tips of the stems. Adult moths are somewhat variable in color and patterns.

CUTWORM

(*Orthosia transparens*: Lepidoptera: Noctuidae)

Characteristics: Young caterpillar grayish-brown with white spotted dorsal bands extending along back. Mature caterpillar more uniformly tan to brown with lighter dorsal bands.

Regional habitat: Dune forest clearings in Oregon.

Comments: This rare caterpillar has unusually striking markings for a cutworm.

BOLETE

(*Leccinum arctostaphyli*: Boletaceae)

Characteristics: Fungus cap convex, brownish to reddish; flesh white, turns bluish-gray upon injury. Stalk thick, solid, fibrous. Pores white. Spores brown.

Regional habitat: Pacific Northwest coast from Washington to California.

Comments: This is an ectomycorrhizal fungus that is connected to the roots of bearberry. It is edible but not tasty.

LEAF GALL APHID

(*Tamalia* sp.: Hemiptera: Aphididae)

Characteristics: Adult various shades of green, becoming darker with age; eyes small, reduced, reddish.

Regional habitat: On bearberry from Washington to California.

Comments: Entire populations develop within leaf galls on bearberry. Once the affected leaf curls and fuses, the aphids are well protected from predators. Several galls can occur on the same leaf. Young galls are green but redden with age.

REINDEER LICHEN

(*Cladonia* sp.: Ascomycota: Cladoniaceae)

Characteristics: Lichen body (thallus) composed of both algal and fungal cells, prostate at first then upright. Erect secondary, shrubby-like thalli bear reproductive bodies (apothecia) at their tips.

Regional habitat: Exposed clearings in shady dune forests, especially in northern California.

Comments: Reindeer lichen often covers large patches of bearberry, sometimes stunting growth of the latter. Here you can see bearberry stems pushing up between reindeer lichen thalli. Reindeer lichen is long-lived and slow growing, absorbing water and mineral nutrients from the air and requiring only a substrate (like the bearberry plant) for support.

GROUND CONE

Parasitic plants are often quite strange and ground cone is no exception. It is not difficult to imagine ground cone as an alien from a distant planet. When it first emerges from the ground it resembles a Douglas-fir cone that has fallen in an upright position, thus the common name. Ground cone lacks any trace of green, has abandoned its independence, and now lives at the expense of nearby salal shrubs. There is no need for chlorophyll since nutrients are conveyed through tube-like outgrowths that attach to the roots of the host plant. Purple leaves and flowers, tightly clustered around the stem, resemble scales. Its microscopic seeds look like sand grains. Scraping away soil at the base of ground cone exposes its underground potato-like tubers, which were eaten by Native Americans.

Little white fly maggots tunnel throughout the soft inner tissues, eating and forming puparia from which small, gray, hairy flies emerge. These flies seem to hold the lease for dwelling inside ground cone stems but it wouldn't surprise me if other insects appear from time to time.

GROUND CONE COMMUNITY Herbivore
Ground cone maggot

GROUND CONE

(*Boschniakia hookeri*: Orobanchaceae)

Characteristics: Parasitic. Yellowish-brown to purplish, lacking chlorophyll. Young plant white but darkens with age. Stem single, thick, cone-like, emerges from subterranean tuber. Leaves scale-like, overlapping. Flowers yellow to purple, cluster on stems. Fruit capsules with minute orange seeds shaped like sand grains.

Regional habitat: Along the Pacific Northwest from Washington to California, but probably wherever salal occurs.

Comments: Ground cone is considered edible but the flavor varies depending on the plant's age and where it grows. It obtains nutrients from the roots of salal.

GROUND CONE MAGGOT

(*Pegomyia* sp.: Diptera: Anthomyiidae)

Characteristics: Larva yellowish, tubular; head narrow with black mouth hooks. Pupa tan, forms within damaged plant. Adult fly light gray; head large; eyes dark; wings shiny, transparent, lacking markings.

Regional habitat: Wherever ground cone occurs from Washington to California.

Comments: Larvae and pupae live in the ground cone stem. Adult flies emerge when the stem deteriorates.

GLORY BUSH

Glory bush has sharp-toothed leaves resembling those of a small holly plant, and they are just as painful when brushed up against. It is amazing that a very fragile caterpillar is able to digest such tough leaves. The light blue flowers attract bumblebees. There are several varieties of glory bush in the Pacific Northwest, but the species seems to be native to central and northern California.

GLORY BUSH COMMUNITY Herbivores
Leaf roller
Bumblebee

GLORY BUSH
(*Ceanothus gloriosus*: Rhamnaceae)
Characteristics: Spreading to erect shrub. Leaves evergreen, opposite, toothed along margins. Flowers blue, borne in clusters. Fruit horned capsules.
Regional habitat: Exposed clearings in dune forests from central Oregon to central California.
Comments: Flowers are visited by bumblebees for pollen and nectar.

LEAF ROLLER

(Lepidoptera: Tortricidae)

Characteristics: Caterpillar uniformly green; thoracic shield black; head black.

Regional habitat: Dune forests from central Oregon to central California.

Comments: Even though glory bush has extremely stiff leaves, caterpillars manage to bend the blades around themselves and then weave the edges together with silk.

FIREWEED

It's not often that you encounter an herbaceous perennial over two meters high, which is the height reached by most fireweed plants within dune forests. Several hundred elegant seeds are revealed when the narrow seed capsule splits open. Silky hairs attached to the tops of the seeds allow them to be easily spread by the wind, just like dandelion seeds. The plants also spread by underground roots.

FIREWEED COMMUNITY
Herbivore
Gall gnat

FIREWEED
(*Chamerion angustifolium*: Onagraceae)
Characteristics: Stems large, straight, up to two meters in height. Leaves large, elongate, alternate, mid-veins white; margins entire but slightly wavy. Flowers pink to reddish; borne in long terminal spikes, also in smaller spikes that emerge from leaf bases.

Regional habitat: Back dunes as well as clearings and borders of dune forests and wetlands.
Comments: Various parts of the plant have been used as salad greens and teas. Gall gnat larvae (Diptera: Cecidomyiidae) live inside curled leaves.

TOOTHWORT

In shady areas with only a small amount of light, you can find the delicate pink flowers of toothwort. Seed development is rapid and seeds mature while some flowers are still blooming on the same stalks. Seedpods have a thick black line on each side, as if someone came along with a black felt-tip pen and marked them. If you compare the nearly circular basal simple leaves with the linear compound upper leaves, it is hard to believe they originate from the same plant.

NUTTALL'S TOOTHWORT

(*Cardamine nuttallii*: Brassicaceae)

Characteristics: Perennial herb. Stem erect, simple. Basal leaves broad, cordate; margins slightly wavy. Stem leaves compound with slender leaflets. Flowers with four pinkish-white petals. Fruit narrow pods (siliques).

Regional habitat: Dune and other forests in Oregon and California.

Comments: The variation in leaf shape on toothwort is curious.

FAIRYBELL

Fairybell can be found in dark, shady woods. Delicate yellow flowers hang down like little lanterns contrasting against large, shiny, dark green leaves. While the small, oblong, bumpy, red berries appear appealing, they are not palatable.

FAIRYBELL

(*Prosartes* (=*Disporum*) *smithii*: Liliaceae)

Characteristics: Perennial herb. Leaves oval, alternate, stalkless, venation parallel. Flowers bell-shaped, cream-colored, usually borne in clusters of two to three near branch tips. Ripe fruit orange berries.

Regional habitat: Dune forests from Washington to California.

Comments: Berries are not considered edible.

FRINGED WILLOW HERB

A little perennial that seems to emerge suddenly from a variety of wet locations is fringed willow herb. Fertilized flowers quickly elongate their seed capsule, even before the attached petals have fallen. Capsules are green at first, then pink, and finally deep red. All stages can be found on the same plant. Leaves are not large, but do conceal small, flattened whiteflies and developing flea beetle larvae that live on the undersides of the leaves. Two insect herbivores have adapted the behavior of entering the plant's tissues for a better chance of survival. The willow herb leaf miner caterpillar squeezes its body between the outer and inner leaf epidermis. The willow herb weevil larva resides in the stems, remaining hidden from view its entire life while slowly destroying the stalks.

FRINGED WILLOW HERB COMMUNITY Herbivores
White fly
Leaf miner
Weevil
Flea beetle

FRINGED WILLOW HERB
(*Epilobium ciliatum*: Onagraceae)
Characteristics: Perennial herb. Leaves slender, slightly toothed. Flowers small, white to reddish; borne along stems; petals deeply notched. Mature fruit narrow, deep-red, vertically-positioned capsules.
Regional habitat: Back dunes, dune forests, clearings, and seepage areas along the Pacific Northwest coast.
Comments: This variable species is widespread and considered native to both North America and Asia.

WHITE FLY

(*Tetraleurodes* sp.: Hemiptera: Aleyrodidae)

Characteristics: Larva minute, flattened, oval-shaped, pale white, partly translucent. Adult small, completely white; wings wide, covered by powdery wax.

Regional habitat: Openings in dune forests from Washington to California.

Comments: Larvae are firmly attached to the undersides of leaves where they feed on plant juices. Pupation occurs within the larval skin on the leaf.

LEAF MINER

(*Mompha* sp.: Lepidoptera: Momphidae)

Characteristics: Larva white to cream-colored; head black; legs short. Adult small, narrow, blackish with white markings; palpi upturned; forewings with rows of upraised scale patches.

Regional habitat: Widespread in the Pacific Northwest.

Comments: This species has a Holarctic distribution. Representatives of this genus specialize on members of the Evening Primrose Family.

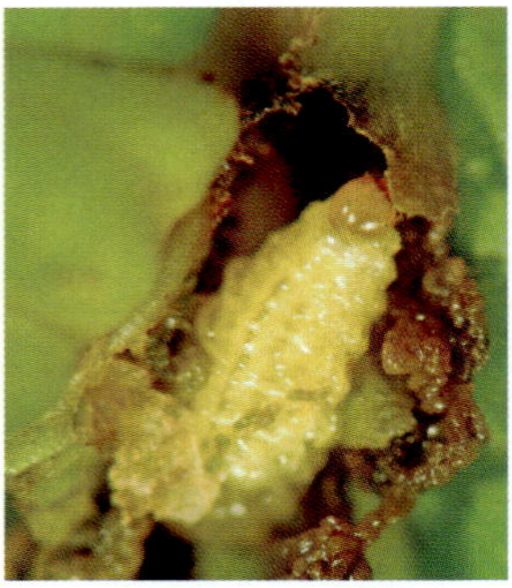

WEEVIL

(*Auleutes epilobii* (=*cruralis*): Coleoptera: Curculionidae)

Characteristics: Larva small, broad, whitish; head tan; legless. Pupa whitish. Adult short, broad, dark brown with tan scale patches covering body; beak long, slightly curved.

Regional habitat: Dune habitats along the Pacific Northwest coast.

Comments: Larvae develop inside stems where they are hidden from view.

FLEA BEETLE

(*Altica foliacea*: Coleoptera: Chrysomelidae)

Characteristics: Larva reddish-brown, covered with rows of black protuberances; legs short. Adult shiny metallic, brownish to black, elytra covered with minute pits; hind legs modified for jumping.

Regional habitat: Coastal dune habitats from Washington to California.

Comments: Larvae are quite conspicuous as they devour the upper leaves of fringed willow herb during the day.

HORSETAILS

Ancestors of today's horsetails, some 300 million years ago, stood 100 meters high, had branches extending at least 3 meters out from their trunks, and dominated the flora of ancient swamps. Large grasshopper-like insects may have fed on these plants, although the silica incorporated into the cell walls certainly would have (and still does) inhibited many potential herbivores. Maybe little sawfly larvae feeding on today's horsetails, that can grow as high as a person's head in dune forests, descended from giant sawflies that fed on prehistoric horsetails.

Some insects attempt to nibble the leaves of modern horsetails from time to time, but only a sawfly, a weevil, and two leaf beetles have developed a steadfast association with these plants. The leaf roller caterpillar munching away on the branches is probably an occasional guest that would prefer to be on a more palatable plant. Females of the weevil and leaf beetles use their sharp mandibles to drill holes into the stems for oviposition. Their developing larvae eat the scanty tissue inside the mostly hollow stems and often finish their development in the rootstalks.

A very strange organism occurs inside horsetail stems. This is a slime mold that can be found by searching for sections of stems that have been twisted lengthwise. In the center of these contorted stems are a number of protruding columnar growths creating an image of an enchanted micro-world. These clear protuberances eventually become covered with a bushy border of spore bodies. Slime molds attract springtails that enter through openings in the stems to feed on the spores.

HORSETAIL COMMUNITY Herbivores	Pathogen	Predator
Sawfly	Slime mold	Spiders
Leaf roller		
Bronze leaf beetle		
Weevil		
Springtail		

HORSETAIL
(*Equisetum* sp.: Equisetaceae)
Characteristics: Perennial, herbaceous. With two types of jointed hollow stems: 1) sterile vegetative with many whorled branches (photo); 2) fertile reproductive, mostly unbranched, bearing persistent terminal cones with spores.

Regional habitat: Moist dune and other forest types throughout the Pacific Northwest coast.
Comments: Horsetails are often considered living fossils; their ancestors occurred in the Paleozoic, some 300 million years ago. They help stabilize the soil structure around the margins of water sources.

LEAF ROLLER

(*Sparganothis senecionana*: Lepidoptera: Tortricidae)

Characteristics: Larva green with white rings separating body segments; head light brownish-green. Adult tan to rust; forewings with V-shape pattern in middle, dark bands at tips.

Regional habitat: Widespread throughout the Pacific Northwest coast.

Comments: Caterpillars web together and live within adjacent leaves of the sterile plants where pupae also are formed. Horsetail is just an occasional host for this polyphagous moth.

SAWFLY

(*Dolerus napaeus*: Hymenoptera: Tenthredinidae)

Characteristics: Larva long, grayish-green; head tan; eyes black. Adult small, completely black; head wide; legs slender; wings membranous, dusky.

Regional habitat: Dune forests from Washington to California.

Comments: Adults are reluctant fliers but scamper rapidly over the surface of horsetail plants. Females have a strongly serrated ovipositor that can penetrate silicified stems and branches. This is one of the few insects that develop only on horsetails.

WEEVIL

(*Grypus equiseti*: Coleoptera: Curculionidae)

Characteristics: Adult short, broad, brown; areas of mottled white scales form bands on sides of prothorax and irregular stripes on elytra; beak long, cylindrical, curved; each elytron bears three elongate tubercles.

Regional habitat: Dune forests from Washington to California.

Comments: This species has a Holarctic distribution and appears to be an obligate feeder of horsetails. Larvae develop inside stems.

BRONZE LEAF BEETLE

(*Hippuriphila equiseti*: Coleoptera: Chrysomelidae)

Characteristics: Adult small, shiny, brownish-black; head reflexed, not visible from above; legs tan.

Regional habitat: Dune forests from Washington to California.

Comments: This species has a Holarctic distribution and appears to be an obligate feeder of horsetails. Larvae develop inside stems.

SLIME MOLD

(*Ceratiomyxa* sp.: Protosteliida: Ceratiomyxidae)

Characteristics: This pathogenic slime mold stimulates cells of sterile horsetail plants to enlarge. The infected area of the stem becomes twisted and deformed. Inside the contorted stem, a naked plasmodium produces whitish, semi-transparent, columnar fructifications.

Regional habitat: Moist areas of dune forests along the Pacific Northwest coast.

Comments: Columnar fructifications of this myxomycete eventually turn opaque and become covered with a mass of tiny sporocarps, each consisting of a stalk supporting a single ovoid spore. Infected horsetail stems break open for spore release and are then invaded by springtails.

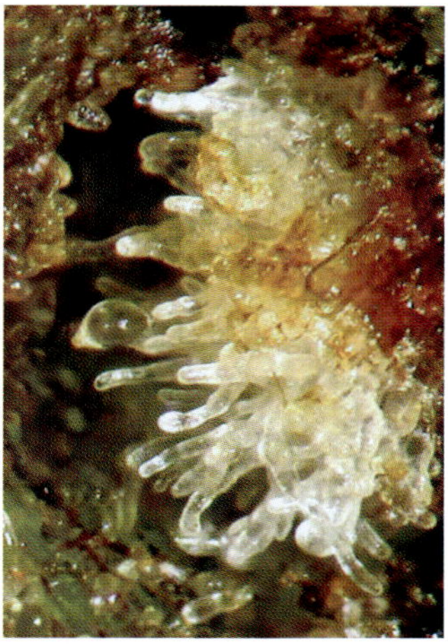

SPRINGTAIL

(*Tomocerus* sp: Collembola: Entomobryidae)

Characteristics: Small, slender wingless hexapod with furcula (jumping organ) under tip of abdomen; antennae long.

Regional habitat: Moist areas of dune forests along the Pacific Northwest coast.

Comments: This species occurs in fractured horsetail stems invaded by slime molds. A series of predators follows, especially spiders. The furcula allows the animal to quickly spring out of danger.

OREGON ASH

Oregon ash often grows in the same habitat as red alder, but its large, compound, opposite leaves easily distinguishes it from the latter. There is an old saying that rattlesnakes will not cross over a twig of Oregon ash. I've been unable to verify this since I never encountered any rattlesnakes in my jaunts through the dune forests. Logs of Oregon ash are occasionally washed up on the beach and, if deposited at the high tide level, may establish root systems and produce small twigs with foliage. These never survive in such a location but this shows the tenacity of Oregon ash.

A wide range of insects, including moths, sawflies, aphids, and gall mites, can be found on Oregon ash. Leaves may be riddled with holes, curled around feeding caterpillars, or peppered with galls. One of the galls had me completely fooled. While dissecting some minute round leaf galls, I found what I thought was a small, elongate egg inside. After placing the "egg" under the microscope, I realized it was a mite; in fact, it is one of the smallest invertebrates known, measuring only a fraction of a millimeter in length. And this mite is not like most that have six legs as nymphs and eight as adults—it has only four legs throughout its entire development and all are positioned at the head end. The small size of these ash mites is perfect for wind dispersal.

OREGON ASH COMMUNITY Herbivores	 Predators-Parasites
Leaf caterpillar	Red-headed soldier beetle
Spiny sawfly	Parasitic wasps
Spotted cutworm	
Pouch gall mite	
Aphids	
Scale insects	
Psyllids	
Yellow-banded cicada	

OREGON ASH

(*Fraxinus latifolia*: Oleaceae)

Characteristics: Medium-sized trees with rough, grayish brown bark. Leaves compound with three to eight smooth, entire leaflets. Flowers small, greenish, appearing at same time as leaves. Fruit samaras.

Regional habitat: Common in dune forests along the Pacific Northwest coast.

Comments: New plants sometimes "sprout" from logs washed up on the beach. Several leaf clusters may appear before the plant perishes.

LEAF CATERPILLAR

(*Oncocnemus* sp.: Lepidoptera: Noctuidae)

Characteristics: Caterpillar light green with several broken white longitudinal lines. Adult light brown; forewings with numerous multicolored patches.

Regional habitat: Coastal dune forests from Washington to California.

Comments: Caterpillars are well camouflaged on the underside of leaves.

SPINY SAWFLY

(*Eupareophora parca*: Hymenoptera: Tenthredinidae)

Characteristics: Young larva greenish above with small black protuberances; sides and undersurfaces yellowish with rows of setal clusters. Mature larva with black above, orange underparts.

Regional habitat: Northwest Pacific coast.

Comments: Body cavities of many larvae contain colonies of yeast. It is not known whether these yeasts are parasitic or symbiotic. Wasp larvae (Hymenoptera) can be found feeding externally on sawfly larvae, eventually killing them.

RED-HEADED SOLDIER BEETLE

(*Podabrus pruinosus*: Coleoptera: Cantharidae)

Characteristics: Adult elongate; head and thorax bright orange-red; eyes dark brown, protruding; antennae long; elytra slender, black, leathery.

Regional habitat: Throughout the Pacific Northwest.

Comments: These carnivorous beetles inspect the leaves of Oregon ash for aphid, psyllid, and scale insect prey.

SPOTTED CUTWORM

(*Lithophane innominata*: Lepidoptera: Noctuidae)

Characteristics: Caterpillar tan, with a wide darker band with white dots running down its back. Adult drab, light gray, with two dark spots on each forewing.

Regional habitat: Washington to California.

Comments: Caterpillars, which also feed on alder and other broadleaf trees, can vary from gray to light green in color.

POUCH GALL MITE

(*Eriophyes* sp.: Trombidiformes: Eriophyidae)

Characteristics: Microscopic, elongate, stout brownish mite with two pairs of anteriorly placed legs.

Regional habitat: Dune forests in Oregon and California.

Comments: Development occurs within small, subspherical, light green pouch galls on ash leaves. These mites, that can transmit plant viruses, are mainly dispersed by wind due to their small size.

YELLOW-BANDED CICADA
(*Okanagana* sp.: Hemiptera: Cicadidae)
Characteristics: Large, slender, dark cicada with a broad head containing protruding eyes; prothorax with a vertical orange mid-line; prothorax separated from mesothorax by a transverse yellow bar; mesothorax with a pair of yellow spots; wings transparent, many-veined, folded roof-like over back at rest.
Regional habitat: Forested areas of the Pacific Northwest.
Comments: Males are capable of making various sounds. Nymphs are subterranean and feed on plant roots, often remaining in the soil for years before emerging.

SALAL

While salal is well adapted to a range of dune habitats and its tough, leathery leaves are unattractive to deer, it does support a wide variety of invertebrates.

A very curious insect that feeds on salal is the primitive timarchian leaf beetle. While a few of these beetles have also been reported from Idaho and Montana, the other 98 species live in Europe. It is not known when or how these North American immigrants arrived. What deepens the mystery is that these beetles are flightless. So did they walk 3,000 miles over the Bering land bridge when it connected Siberia to Alaska long ago, or were they carried unnoticed by early humans coming to America?

Today, these beetles are difficult to find along the Pacific Northwest coast and after studying them, I was amazed that any had survived at all. Plump beetle larvae that remain exposed during the day and can easily be hand-picked from the leaves would appear to be quite vulnerable. The main line of defense of the adults when disturbed is to drop like lead weights, since they can't run or fly.

The sedentary larvae are susceptable to diseases and parasites. Small flies lay their eggs on the surface of these larvae. Upon hatching, the maggots puncture the larval skin and feed on exposed blood. Since the larvae appear to have no defense against this insult, they stoically tolerate it until the fly is finished feeding or the larva itself dies. Beetle larvae fortunate enough to escape the flies then enter the soil to pupate, but here they may fall prey to small microscopic nematodes that turn their bodies into a soupy mass.

Walking through a patch of salal in summer can be a treat when the delicate little white moths have emerged and are flittering from plant to plant in search of mates. Most curious is a tiny black spot in the center of each forewing. White moth caterpillars are reclusive, and either bury themselves in salal stems or cover themselves with a matting of leaves webbed together with silk. The white moths are susceptable to protozoan parasites that turn their blood a pinkish color.

It's amazing how so many different caterpillars with different life-styles feed on salal. Besides the white moth caterpillars, there is a leaf roller that lives inside a tightly curled leaf, only coming out to feed at night. Then there is a very small leaf mining caterpillar that lives between the upper and lower surfaces of the leaves, feeding continuously on the tender internal tissue and forming a weakened area in the cuticle for its emergence after pupation. These caterpillars are easy to find since their blotch mines leave large brown patches on the leaves, much larger than the brown spots caused by a leaf spot fungus. Then there are looper caterpillars that feed completely exposed on the leaves, but escape notice by camouflaging themselves to appear as short stems.

Most never see the salal spittlebug nymph because it is entirely covered with spittle. What an amazing and very successful habit: imbibing plant juices at one end and excreting them as soap bubbles at the other end. While not actually soap, the bubbles have qualities similar to liquid detergents.

SALAL COMMUNITY Herbivores	Predators-Parasites	Omnivores
Leaf beetle	Ladybird beetle	Small milkweed bug
White moth	Soldier beetle	Banana slug
Tip borer	Leaf spot fungus	
Leaf blotch miner	Nematodes	
Leaf roller	Fly larvae	
Aphids	Microsporidia	
Soft scale		
Spittlebug		
Stem maggot		
Hairy-shelled snail		
Gall midge		
Looper		

SALAL
(*Gautheria shallon*: Ericaceae)
Characteristics: Medium to tall evergreen shrub. Leaves thick, leathery, with finely serrated margins. Flowers white to pinkish, bell-shaped; borne in short, one-sided clusters. Fruit berry-like, blue when mature.

Regional habitat: Dune forests along the Pacific Northwest.
Comments: Mature fruits are edible and used locally in jams and pies. Salal often forms dense impenetrable thickets. Shallon is another common name for salal.

LEAF BEETLE

(*Timarcha intricata*: Coleoptera: Chrysomelidae)

Characteristics: Egg elongate, deposited inside a brown ball of fecal material. Larva light brown, hemispherical in side view; head dark brown. Pupa light brown, short, wide. Adult stocky, shiny, reddish to black; flightless.

Regional habitat: Dune forests from Washington to California.

Comments: Larvae, which also feed on coast rhododendron and evergreen huckleberry, are completely exposed during the day. While their blood is considered to contain toxic compounds, they are still attacked by fly predators and nematode parasites.

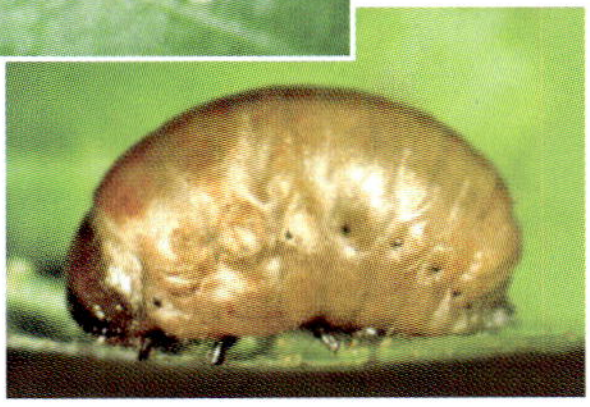

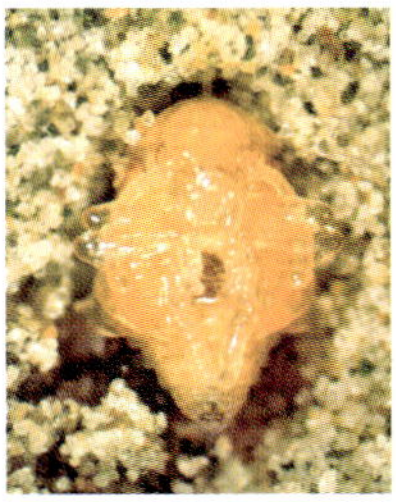

LADYBIRD BEETLE

(*Mulsantina picta*: Coleoptera: Coccinellidae)

Characteristics: Larva elongate, dark gray with white lines down middle of back and along both sides; head and legs brown. Adult oval; elytra pale with brown band along edges and across middle and toward tip.

Regional habitat: Dune forests from Washington to California.

Comments: This predaceous species is common on salal where it searches for its favorite meal, aphids.

WHITE MOTH

(*Eucalantica* sp.: Lepidoptera: Yponomeutidae)

Characteristics: Caterpillar stubby, whitish, with brown head. Adult completely white except for a small black dot in each forewing.

Regional habitat: Dune forests from Washington to California.

Comments: The white adults are quite striking when they fly *en masse* among the salal plants. Similar caterpillars also attack coast rhododendron. Rod-shaped microsporidian parasites (protozoa) infect the caterpillars, turning their bodies various shades of pink.

TIP BORER

(*Epinotia* sp.: Lepidoptera: Tortricidae)

Characteristics: Caterpillar white, body with a few straight setae; head brown. Adult grayish-brown; forewings with dark band across upper portion when closed.

Regional habitat: Dune forests in Washington and Oregon.

Comments: Feeding activity of caterpillars inside the salal stem usually destroys the growing tip.

LEAF BLOTCH MINER

(*Cameraria gaultheriella*: Lepidoptera: Gracillariidae)

Characteristics: Caterpillar minute, straight, pale yellow; head pointed; body segments distinct, with dark, broken, dorsal band. Adult with orange-brown scaly forewings bearing several angled white bands.

Regional habitat: Dune forests from Washington to California.

Comments: Caterpillars live in blotch mines between the upper and lower leaf epidermis. Similar leaf blotch miners occur on twinberry and evergreen huckleberry.

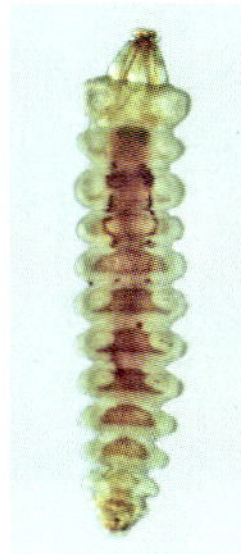

LEAF ROLLER

(*Acleris* sp.: Lepidoptera: Tortricidae)

Characteristics: Caterpillar body and head orange-brown. Pupa slender, brown. Adult orange-brown; forewings with diffuse broken dark band across middle.

Regional habitat: Dune forests in Oregon.

Comments: Larvae fold over leaf edges and feed within a silken web.

SOFT SCALE

(*Eulecanium* sp.: Hemiptera: Coccidae)
Characteristics: Females domed, reddish-brown; dorsal and lateral surfaces with rows of irregular white bands.
Regional habitat: Dune forests from Washington to California.
Comments: Sedentary females appear as large "warts" on the stems and leaves of salal.

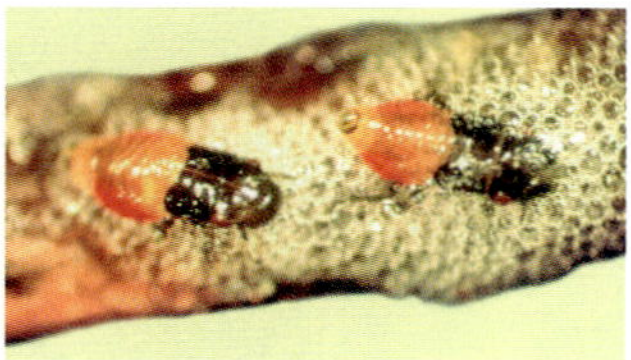

SPITTLEBUG

(*Aphrophora* sp.: Hemiptera: Aphrophoridae)
Characteristics: Nymph small; abdomen reddish; thorax and head black. Older nymphs have different color configurations with large bluntly rounded heads. Reddish eyes remain constant.
Regional habitat: Dune forests throughout the Pacific Northwest.
Comments: Nymphs feed on juices of salal while protected by a mass of "spittle."

SOLDIER BEETLE

(*Podabrus* sp.: Coleoptera: Cantharidae)
Characteristics: Adult elongate, reddish-brown; eyes bulging; antennae long; prothorax narrow.
Regional habitat: Dune forests from Washington to California.
Comments: Members of this genus are predators that search for small prey, especially aphids, spittlebugs, and scale insects.

SMALL MILKWEED BUG

(*Lygaeus kalmii*: Hemiptera: Lygaeidae)

Characteristics: Body mostly black but thorax with red bar or spots; wing bases with red cross bars; wing membranes black, often with pair of white spots.

Regional habitat: Dune forests in Oregon.

Comments: While the common name implies an association with milkweed, these bugs occur on a wide range of plants, including salal. They are omnivores, feeding on both plant and animal matter.

STEM MAGGOT

(*Pegomyia* sp. Diptera: Anthomyiidae)

Characteristics: Maggot light yellow; small protuberances on body; mouth hooks black. Puparium elliptical, brown or red, formed inside stem. Adult small, blue-gray; wings transparent.

Regional habitat: Dune forests in Washington and Oregon.

Comments: Maggots feed within stems, often killing stem tips. Puparia darken with age.

HAIRY-SHELLED SNAIL

(*Vespericola columbianus*: Gastropoda: Polygyridae)

Characteristics: Flattened many-coiled brown shell covered with numerous fine, short setae.

Regional habitat: Dune forests from Washington to California.

Comments: These snails only occur along the Pacific coast. Other snail types with small conical shells, such as those of the family Pupillidae (bottom photo), also feed on salal leaves.

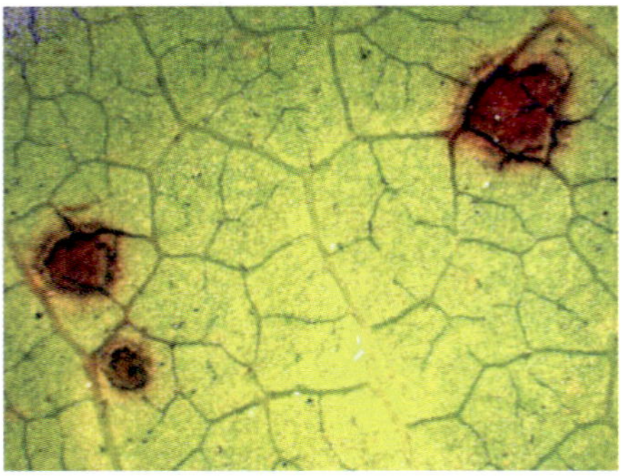

LEAF SPOT FUNGUS
(*Phyllosticta* sp.: Ascomycota: Botryosphaeriaceae)
Characteristics: Fungus produces small, irregular, reddish-brown spots on leaves. Black fruiting bodies (pycnidia) eventually develop in center of these spots.
Regional habitat: Dune forests from Washington to California.
Comments: The fungus can develop on stems as well as leaves of salal.

GALL MIDGE
(*Dasineura* sp.: Diptera: Cecidomyiidae)
Characteristics: Maggot small, robust, light tan; head pointed; mouth hooks black. Puparium reddish-brown. Adult small, black; antennae long.
Regional habitat: Dune forests from Washington to California.
Comments: Maggots live and feed inside curled leaves. Pupation can occur in the larval feeding site or in the ground. Other species of gall gnats form large tuberous galls filled with numerous larvae on salal stems.

LOOPER
(*Thallophaga* sp.: Lepidoptera: Geometridae)
Characteristics: Caterpillar tan with pale line extending along side of body; with two pairs of posterior pseudolegs. Adult grayish; forewings with dark pepper spots.
Regional habitat: Dune forests from Washington to California.
Comments: A related species of this leaf-eating moth occurs on sword fern.

BANANA SLUG

(*Ariolimax columbianus*: Gastropoda: Ariolimacidae)

Characteristics: Large, yellowish to gray slug usually bearing irregular dark spots.

Regional habitat: Dune forests from Washington to California.

Comments: Banana slugs, which feed on both living and dead plant matter, including mushrooms, are only found in dune forests along the Pacific Northwest coast. This is the largest American slug, with members reaching up to 230 mm in length.

LARGE-LEAVED AVENS

Deep in the shady woods occurs yet another, usually solitary, perennial, the large-leaved avens. Dark green leaves contrasting with small, yellow, timid-appearing flowers is an identifying characteristic. In spite of the luscious appearing leaves, no insects could be found on this plant. The leaves have been used as an astringent.

LARGE-LEAVED AVENS
(*Geum macrophyllum*: Rosaceae)
Characteristics: Hairy perennial. Leaves lobed, serrate. Flowers small, yellow; borne singly or in small clusters at stem tips. Seeds with hairs for attachment on passing animals.

Regional habitat: Dune forests from Washington to California.
Comments: Large-leaved avens grows in shady areas of the forest together with toothwort and fairybells.

GOLDEN ASTER

In sunny clearings can be found groups of golden asters with their yellow flowers reaching for the light. Many of the heads seemed to be drying up prematurely and a closer examination showed that they were all infested by one or more maggots that had eaten out the flower bases. Both maggots and puparia could be found in the same damaged flower. Deeper in the flower receptacles occurred small gall midge larvae. It is interesting to see how insects often partition their food source. In this case, head fly maggots feed on upper florets while the gall midge young develop on seeds.

GOLDEN ASTER COMMUNITY Herbivores
Head fly
Gall midge

GOLDEN ASTER
(*Heterotheca sessiliflora*: Asteraceae)
Characteristics: Perennial. Leaves small, elliptical, entire. Flower heads large; long yellow ray (outer) petals; short disk (inner) petals. Seed heads fluffy.

Regional habitat: Dune forests in California.

Comments: This plant often forms dense mats in open clearings. Populations may also extend down into the back sand dunes. Dandelion-like seeds are spread by wind.

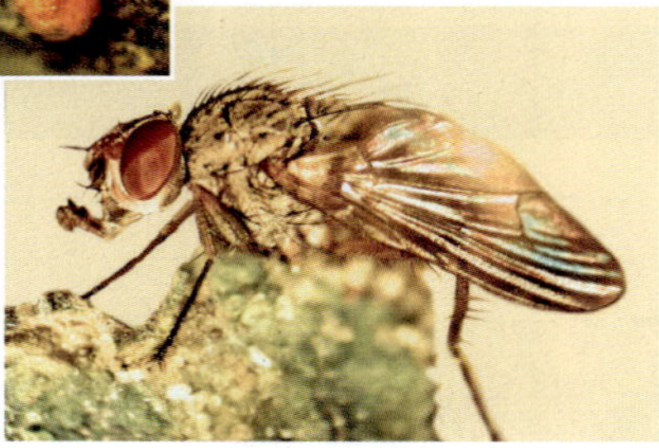

HEAD FLY

(*Pegomyia* sp.: Diptera: Anthomyiidae)

Characteristics: Maggots white to gray, legless, with black mouth hooks. Pupa elliptical, brown, formed within flowerhead. Adult grayish; eyes large, brown; wings membranous, metallic.

Regional habitat: Dune forests in California.

Comments: The head fly maggots devour the developing florets while the minute reddish gall midge maggots (middle photo) concentrate on the seeds of golden aster.

OCEAN SPRAY

A tall perennial shrub with triangular-shaped inflorescences containing numerous small white flowers is likely to be ocean spray, one of the more showy plants of dune forests. Flowers remain on the plant long after they have served their purpose, and that is fortunate for noctuid caterpillars that live in the dried flower clusters. Lorquin's admiral caterpillar, whose back is camouflaged with patches of different shades of green, develops on the leaves. Wings of Lorquin's admiral butterfly are especially striking with a row of white patches outlined against a black background.

While walking through the woods during a summer evening, I would frequently here the drumming of a katydid and immediately know that there was an ocean spray shrub in the vicinity. Not that drumming katydids are restricted to this particular plant, but they do tend to suddenly appear when ocean spray flowers bloom. When quiet, katydids are hard to detect since they remain completely motionless and blend in against the leaves. The female's large, saber-shaped ovipositor seems out of proportion with the rest of her body. It is used to place eggs deep within the stem of the host tree. Katydids don't seem to mind if they ingest a few insects along with the leaves and are considered omnivores.

OCEAN SPRAY COMMUNITY Herbivores	Omnivore
Lorquin's admiral	Drumming katydid
Cutworm	

OCEAN SPRAY

(*Holodiscus discolor*: Rosaceae)

Characteristics: Tall shrub. Leaves broad, toothed, alternate. Flowers white to cream; borne in triangular-shaped dense terminal clusters. Fruit small, dry achenes.

Regional habitat: Dune forests along the Pacific Northwest from Washington to California.

Comments: The hard wood of ocean spray was used by Amerindians for spears, bows, arrows, and harpoons. The fruit was used for medicinal purposes.

LORQUIN'S ADMIRAL

(*Limenitis lorquini*: Nymphalidae)

Characteristics: Caterpillar greenish with a white stripe along its sides; some segments with dorsal protrusions. Adult mostly black; all wings bearing broad white bands; forewings with orange tips.

Regional habitat: Dune forests from Washington to California.

Comments: Caterpillars are quite striking with their dorsal protrusions. The adult wing pattern is remarkable.

CUTWORM

(*Lithophane* sp.: Lepidoptera: Noctuidae)

Characteristics: Caterpillar brownish-gray; three narrow white lines running length of body; head and pronotum brown.

Regional habitat: Dune forests from Washington to California.

Comments: Caterpillars occur within flower clusters.

DRUMMING KATYDID

(*Meconema thalassinum*: Orthoptera: Tettigoniidae)

Characteristics: Grasshopper-like, medium to large, green; legs spiny; antennae long; forewings reticulate. Female has long, robust, saber-shaped ovipositor.

Regional habitat: Dune forests in Washington and Oregon.

Comments: Drumming katydids spend most of their lives on shrubs and trees where they consume plant material as well as small insects. The drumming is produced by vibrations when the katydids rub their hind legs against leaves.

MARSH PENNYWORT

The shape and venation of the marsh pennywort leaf are quite noticeable. Here is a plant with leaves round and symmetrical, and the petiole attached perfectly in the center of the blade. The small white flowers seem out of proportion with the large leaf. A green leaf roller caterpillar easily bends one edge of a leaf over to make a secure compartment for itself. Rusty brown leaf roller moths can usually be found in the vicinity of marsh pennywort.

MARSH PENNYWORT COMMUNITY Herbivore
Leaf roller

MARSH PENNYWORT
(*Hydrocotyle verticillata*: Araliaceae)
Characteristics: Creeping perennial, rooting from stem nodes. Leaves round, crenulate; petiole attached in middle. Flowers minute, white, in small clusters at tips of branchlets. Fruit flat, roundish berries.

Regional habitat: Widespread in damp areas in dune forests from Washington to California.
Comments: It is unusual to have leaves with petioles attached to the center of the blades.

LEAF ROLLER

(Lepidoptera: Tortricidae)

Characteristics: Caterpillar green with small dorsal protuberances; head brown. Adult uniformly reddish-brown; forewings with row of darker scales along tips.

Regional habitat: Dune forests in Oregon.

Comments: Caterpillars fold over the leaf and spin a web between the leaf blade and petiole.

PACIFIC WATERLEAF

The most striking feature of Pacific waterleaf are the long purple stamens that extend far above the flower head, a condition often found in plants that are wind pollinated. Maple-like leaves provide food for a large weevil that is perfectly colored to avoid detection. Its grayish-green body, covered with a series of small black dots, matches the leaves. Occasionally, this weevil is found feeding on strawberry leaves, but is rarely a pest. Weevil larvae were not found.

PACIFIC WATERLEAF COMMUNITY Herbivore
Waterleaf weevil

PACIFIC WATERLEAF
(*Hydrophyllum tenuipes*: Boraginaceae)
Characteristics: Erect perennial herb. Leaves lobed, opposite. Flowers small, lavender to white; borne in tight clusters at stem tips; stamens and pistils extend beyond flower head. Fruit is a capsule.

Regional habitat: Moist, shaded areas of dune forests along the Pacific Northwest coast from Washington to California.

Comments: Leaves and roots of Pacific waterleaf are considered edible.

WATERLEAF WEEVIL

(*Panscopus gemmatus*: Coleoptera: Curculionidae)

Characteristics: Adult large, grayish-brown; thorax and elytra with broken rows of black dots; snout straight; antennae elongate; snout and antennae similar in color.

Regional habitat: Moist, shaded dune forests in Washington and Oregon.

Comments: Adults feed on leaves of waterleaf. Larvae were not found.

KLAMATHWEED

Sometimes, plants from foreign lands quickly adapt to new conditions and sweep across the nation from coast to coast, often crowding out native species. Klamathweed, also called St. John's wort, is a good example. While admiring the yellow flowers of this plant in a dune forest clearing, I remembered that they are not as innocent as they appear and that, in the past 100 years, large funds have been spent to control fields of Klamathweed in California. Klamathweed contains the oil, hypericin, that, when ingested, makes animals sensitive to sunlight and can result in rashes, blisters, and even death if eaten in excess.

Due to Klamathweed's wide range, it was decided to use biological control to reduce its further spread, The theory of biological control is fairly simple. Find out what is keeping the plants in check in their native land and then bring those agents over and release them in the recently infested country. The goal is to re-establish a balance between the plant and its herbivores in the new domain. This involves much diligent preliminary work to be certain that the agents selected for release will not become pests themselves on native, ornamental, or food crops.

Among the agents released to halt expanding populations of Klamathweed in the Pacific Northwest were two leaf beetles acquired from abroad. They quicky adapted to North American conditions and within 30 years, populations of Klamathweed had been significantly reduced. Both species of these leaf beetles species can be found on Klamathweed plants in dune forests.

KLAMATHWEED COMMUNITY Herbivores	 Parasite
Green leaf beetle	Bug
Black leaf beetle	
Leaf roller	

KLAMATHWEED

(*Hypericum perforatum*: Hypericaceae)

Characteristics: Perennial herb. Stems erect. Leaves opposite, sessile, with translucent dots (seen best when held against the light) and black specks. Flowers bright yellow; borne in clusters at stem tips.

Regional habitat: Widespread in dune forests along the Pacific Northwest coast.

Comments: After its introduction, Klamathweed, also known as St. John's wort, became a noxious and invasive weed in many areas of North America. The importation and establishment of Klamathweed-feeding insects from abroad greatly reduced populations of this plant.

GREEN LEAF BEETLE
(*Chrysolina hyperici*: Coleoptera: Chrysomelidae)
Characteristics: Larva plump, hump-backed, grayish with white spots; head dark brown to black. Pupa yellowish-orange; eyes dark reddish, more evident toward end of development. Adult greenish-bronze, round, elytra with broken rows of small pits.
Regional habitat: Dune forests from Washington to California.
Comments: This species and the black leaf beetle (*C. quadrigemina*) (bottom photo) were introduced into North America as biological control agents of Klamathweed. Larvae and adults of both species feed on foliage and, when numerous, can destroy the plant.

LEAF ROLLER
(*Acleris* sp.: Lepidoptera: Tortricidae)
Characteristics: Caterpillar with pale green head capsule with a few brown markings. Adult light reddish-brown; forewings with dark brown lines, wing tips with brown margins.
Regional habitat: Dune forests in Oregon.
Comments: The caterpillar of this moth, which binds together and develops inside Klamathweed leaves, is probably a general feeder. A small immature bug (Hemiptera) was noted with its stylet inserted into a pupa, obviously feeding on body fluids.

BLUE FLAG

The structure of the flower of blue flag and other iris species is complex. Petals (called standards) are in a vertical position and only the undersides are visible. The objects with white lines (called signals) that emerge from the base are outstretched purplish sepals (called falls). Undoubtedly, these white lines help guide insects searching for nectar and pollen to the center of the flower where the anthers and stigma are located. Bumblebees seem to read these signals very well as they are frequent visitors to blue flag flowers.

BLUE FLAG
(*Iris tenax*: Iridaceae)
Characteristics: Perennial herb. Leaves long, narrow, basal. Flowers large, blue to purple, composed of three outstretched sepals and three upright petals approximately equal in size; borne at stem tips.

Regional habitat: Open spaces in dune forests from Washington to California.
Comments: Blue flag occurs in several habitats, especially near moist clearings in and around the edges of dune forests. Flowers are visited by bumblebees.

LICHENS

Lichens are the epitome of comradeship: two plants existing in concert with a common aim, survival. The close association of a fungus and an alga allows lichens to survive under extreme physical conditions that would be lethal to other plants and to each lichen component if they lived separately. The alga (or phycobiont) contains green clorophyll but is delicate and easily killed by desiccation. The fungus (or mycobiont) lacks chlorophyll but provides a protective layer around the alga and obtains food from it. This is an excellent example of symbiosis, where two organisms live in close physical contact. More specifically, they exhibit mutualism, where both benefit from the relationship.

Along the Pacific Northwest coast, marine wart lichens are well known for surviving under extremely harsh conditions. Growing completely exposed on the surface of large boulders along the seashore, wart lichens tolerate pounding waves with each incoming tide. They are so well anchored they are impossible to remove without fracturing the underlying rock.

Most lichens reproduce asexually by forming small spores containing cells of both the alga and the fungus. However, the fungal portion of many dune forest lichens sexually reproduce by producing mushrooms. Few observers realize that fungal strands at the base of many delicate mushrooms are connected to globular green masses of algal cells.

Lichens occur in all shapes and sizes and some of my favorites are the little club lichens. Rarely more than 50 mm in height, club lichens often grow in groups and, while all are fairly straight without any branches, the tips of different species vary in shape. Pixie-cup lichens have an open funnel-shaped tip while chimney-club lichens have narrow-rimmed tops that resemble miniature factory smokestacks. Those with tops bearing red fruiting bodies are known as lipstick lichens. Then there are leaf lichens that emerge between clumps of moss on the forest floor. Their spreading, flat, bi-colored thalli (growing parts) resemble the fruiting bodies of some pore mushrooms.

While many lichens are nearly too small to notice, you can't help but see the large, many-branched hair or bearded lichens hanging from tree branches in dune forests. In winter, when alders and other trees have lost their leaves, the now-exposed hair lichens are quite conspicuous and often referred to as winter foliage. Kids use beard lichens to perk up their masks at Halloween, not realizing that these mats are filled with small invertebrates such as tardigrades and nematodes. Since lichens dry out and become brittle in the summer, what happens to these creatures? Surprisingly, these miniature animals will also dry out and remain in suspended animation until winter rains return and bring them back to life. If you take dry hair lichens in the middle of summer and place them in a bowl of water overnight, you will probably find little wiggling organisms that came out of their summer sleep.

Lungwort lichens represent a different type of leaf lichen, one that grows up to a foot across on tree limbs. In spite of rather insecure attachment, lungwort lichens manage to remain in place throughout most winter storms. They have wide, green lobes covered with shallow grooves that give the appearance of lungs. Curiously, at one time they were used to treat lung disorders. Today they are used as an antiseptic and to treat asthma. While filled with a number of chemicals that are thought to protect against invertebrate herbivores, lungwort lichens are sensitive to air pollution and could be used as an environmental quality sensor. To show how intricate and complex lichens can be, it was recently discovered that the lungwort lichen is not only a partnership between a fungus and alga but also a cyanobacterium. So we have a case of symbiosis involving organisms from three different kingdoms!

WART LICHEN
(*Staurothele fissa*: Ascomycota: Verrucariaceae)
Characteristics: Marine lichen that forms crust on rock surfaces. Thalli uneven, dark greenish-brown to brownish-black. Fruiting structures (perithecia) partially immersed in thalli.
Regional habitat: On rocks all along the Pacific Northwest coast.
Comments: Its thalli are so tightly attached to boulders that wart lichen cannot be removed without chipping the rock.

FALSE PIXIE CUP
(*Cladonia chlorophaea*: Ascomycota: Cladoniaceae)
Characteristics: Thalli erect, unbranched; greenish clubs (podetia) expand into cups at tips.
Regional habitat: On humus and tree trunks in dune forests throughout the Pacific Northwest.
Comments: This is one of several cup lichens found in dune forests.

CHIMNEY-CLUB LICHEN

(*Cladonia* sp.: Ascomycota: Cladoniaceae)

Characteristics: Thalli erect, unbranched; greenish clubs (podetia) with small, expanded blunt tips.

Regional habitat: On humus and tree trunks in dune forests throughout the Pacific Northwest.

Comments: This is one of several club lichens that grows in dune forests.

DOG PELT LICHEN

(*Peltigera membranacea*: Ascomycota: Peltigeraceae)

Characteristics: Thalli flat, spreading, dark-gray with light gray margins. Fruiting structures small orange to red bodies along thalli edges.

Regional habitat: Widespread in mossy areas in dune forests in the Pacific Northwest.

Comments: Snake fly larvae (*Agulla* sp.: Raphidioptera; Raphidiidae) patrol the surface of leaf lichens, probably preying on other small arthropods living in the habitat.

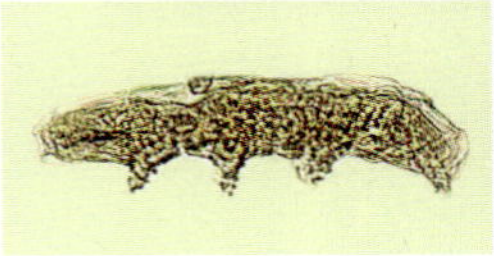

BEARDED LICHEN

(*Usnea* sp.: Ascomycota: Parmeliaceae)

Characteristics: Hanging, hairy lichen; stems many-branched, curved.

Regional habitat: Hanging on trunks and branches of trees in dune forests in the Pacific Northwest.

Comments: Bearded lichens are especially noticeable on broadleaf trees that have lost their leaves. Stubby, eight-legged tardigrades (bottom left) and long, slender nematodes (bottom right) remain on bearded lichens all year, entering a dormant, desiccated stage along with the lichen during dry summer months and becoming active when the rains return.

LUNGWORT LEAF LICHEN

(*Lobaria pulmonaria*: Ascomycota: Lobariaceae)

Characteristics: Large, dangling, leafy lichen with broad lobes. Thalli green on upper surface, brownish beneath. Asexual reproductive structures (soridia and isidia) on ridges and margins of thalli.
Regional habitat: On trees in dune forests from Washington to California.
Comments: At one time, lungwort was used as a treatment for pneumonia. It is sensitive to air pollution. The absence of herbivores may be due to the arsenal of chemicals found in lungwort lichen tissues.

LICHEN MUSHROOM

(*Lichenomphalia umbellifera*: Agaricales: Hygrophoraceae)

Characteristics: Alga comprises a dark green, globular mass embedded in mucilage surrounded by a thick envelope of fungal mycelia that forms a mushroom (*Omphalina ericetorum*). Mushroom cap flattened- to funnel-shaped, pale buff to brown, often with faint bands radiating out from center to margins like spokes in a wheel.
Regional habitat: Dune forests throughout the Pacific Northwest.
Comments: The lichen mushroom (basidiolichen) is the fruiting body of the fungus that forms the fungal-algal association.

TIGER LILY

The orange flowers of the tall tiger lily are quite striking against the dark green background of dune forests. It is possible that the spots on the sepals and petals serve to attract pollinators.

TIGER LILY
(*Lilium columbianum*: Lilaceae)
Characteristics: Tall perennial. Leaves narrow, pointed, arranged in whorls along single straight stem. Flowers large, bent downwards; petals and sepals similar, yellow with reddish spots.

Regional habitat: Dune forests from Washington to California.
Comments: Bumblebees and butterflies visit the large showy flowers. Amerindians cooked and ate the bulbs.

TWINFLOWER

This little evergreen is not easy to find since it is typically never more than 50 mm in height and the trailing stems are usually hidden by leaves of neighboring plants. A pair of trumpet-shaped flowers are borne on the same stalk. The flowers are said to be quite fragrant. Small fruits are sticky and attach to hair and feathers of passing animals. This was one of the favorite flowers of the Swedish naturalist, Carl Linnaeus, and the scientific name of the genus, *Linnaea*, reflects this.

TWINFLOWER
(*Linnaea borealis*: Caprifoliaceae)
Characteristics: Small creeping perennial. Leaves small, oval, shiny, opposite. Flowers whitish-pink, pendulous; borne in pairs at tips of leafless stems. Fruit sticky nutlets that adhere to pelage of animals.

Regional habitat: Dune forests along the Pacific Northwest coast from Washington to California.
Comments: Since plants are self-incompatible, local spread is usually by means of stolons. Dispersal is by sticky seeds.

TWINBERRY

Twinberry is a honeysuckle bush with paired yellow flowers that obviously bear a good supply of nectar based on the number of bees and hummingbirds that visit them. Its two black berries surrounded by reddish bracts provide the plant's name. The berries were appealing to my dog who would spend several minutes nuzzling and licking them but would never eat any. I quickly discovered why since they are bitter with a strong chemical taste. Native Americans called the berries "raven's food," but I never saw ravens or crows sampling any. Apparently twinberry juice was used to dye hair. Bark of twinberry supposedly has some medicinal properties. Striking speckled sawfly larvae commonly feed on twinberry leaves.

There is an interesting case of symbiosis that occurs on twinberry leaves. First, a small fly lays its eggs inside leaf tissue. The leaf responds by swelling up and forming a gall around the hatching eggs and developing maggots. Soon a different type of fly arrives and lays its eggs on the leaf gall. Then moths arrive to deposit their eggs on the gall, and eventually a mixture of fly maggots and caterpillars are all sharing the same gall. How they are able to tolerate each other is a mystery.

TWINBERRY COMMUNITY Herbivores
Sawfly
Stem fly
Leaf roll gall midge
Plume moth
Tissue moth
Leaf moth
Leaf blotch miner

TWINBERRY

(*Lonicera involucrata*: Caprifoliaceae)

Characteristics: Tall straggly shrub. Leaves opposite, broadly elliptical, pointed. Flowers yellow; borne in pairs from leaf axils. Fruit shiny black, subtended by large reddish bracts.

Regional habitat: Dune forests from Washington to California.

Comments: Fruits are not palatable but were used by Native Americans as a black pigment for various purposes, including dyeing hair. The plant is also known as bearberry honeysuckle.

SAWFLY

(*Abia americana*: Cimbicidae)

Characteristics: Larva light to dark gray; body with transverse rows of black and yellow spots; head brown. Pupa whitish, formed inside brown cocoon. Adult black, hairy; antennae with terminal segments enlarged (clubbed); wings membranous with brown blotches.

Regional habitat: Dune forests from Washington to California.

Comments: Larvae darken as they grow and the yellowish spots become more evident. Another host plant for this sawfly is the Himalayan honeysuckle (*Leycesteria formosa*).

STEM FLY

(*Pegomyia* sp.: Diptera: Anthomyiidae)

Characteristics: Maggot white, slender; mouth hooks black. Puparium brown. Adult gray; head brown; eyes red; wings membranous.

Regional habitat: Dune forests in Oregon.

Comments: Larvae develop inside the stem of twinberry and pupate in the ground or inside the hollowed-out stems. The adult emerges head first from one end of the puparium.

LEAF ROLL GALL MIDGE

(*Contarinia* sp.: Diptera: Cecidomyiidae)

Characteristics: Larva very small, white to yellowish; mouth hooks black. Pupa brown. Adult small; black; wings clear.

Regional habitat: Dune forests in Oregon.

Comments: The succulent leaf roll gall is quite moist inside, so much so that the maggots could almost be considered semi-aquatic. Up to 12 maggots can occur inside a single gall.

PLUME MOTH

(*Amblyptilia pica*: Lepidoptera: Pterophoridae)

Characteristics: Caterpillar mottled gray. Adult light-brown with darker markings; with long legs and long narrow wings.

Regional habitat: Dune forests in Oregon.

Comments: Caterpillars are frequently found in leaf galls made by the leaf roll gall midge.

TISSUE MOTH

(*Triphosa haesitata*: Lepidoptera: Geometridae)

Characteristics: Caterpillar brownish with small faint yellow spots; head tan. Adult brown; forewings with irregular brown lines; fore- and hindwings with scalloped edges on tips.

Regional habitat: Dune forests from Washington to California.

Comments: Caterpillars feed on galls made by twinberry leaf roll gall midges. Coloration of the adult moths varies greatly.

LEAF MOTH

(*Euceratia securella*: Lepidoptera: Ypsolophidae)

Characteristics: Caterpillar slender, greenish-yellow, with mid-section swollen; terminal segment bifid; head small. Pupa slender, brown; formed in cocoon shaped like a circus tent. Adult brownish; palps and antennae often outstretched at rest; forewings with dark and light markings, tips bear erect scale tufts.

Regional habitat: Dune forests from Washington to California.

Comments: Caterpillars occur on leaves of twinberry as well as in leaf roll galls.

LEAF BLOTCH MINER

(*Cameraria* sp.: Lepidoptera: Gracillariidae)

Characteristics: Caterpillar small, pale yellow; segments distinct; head partially withdrawn into thorax. Pupa brown, slender; formed in silken cocoon. Adult small; forewings with beautiful color pattern of irregular orange, white, and black bands.

Regional habitat: Dune forests in Washington and Oregon.

Comments: This colorful little moth is similar to species on salal and evergreen huckleberry.

LIVERWORTS

While walking through a dune forest or any forest for that matter, it is usually larger plants that are noticed while smaller, more inconspicuous ones blend into the landscape. In this latter group are petite, miniature mosses and liverworts that cover soil, rocks, and fallen logs.

Liverworts (hepatics) and mosses (bryophytes) are simple plants with reduced vascular tissue (vessels that carry water, sap, and nutrients). Leafy liverworts look like mosses but their opposite leaves lack a central midrib and they have dark capsules. In fact, it is the capsules on leafy liverworts that are so fascinating. Borne on erect stalks, the capsules remind me of a series of minuscule matches that were lighted then immediately blown out and stuck in the ground. These capsules are filled with spores that obviously are nutritious, at least to the beetle larvae that devour them. Most curious are crane fly larvae feeding on leafy liverworts growing on tree limbs since crane flies usually develop in decaying wood or in the soil. During my investigation of dune fauna, I only found exposed crane fly larvae twice, in liverworts and on leaves of snakeberry.

A second group of liverworts are the thalloid types, one of the more common species being lung liverwort. In this plant, the "leaves" are flattened and lobe-like. The sexes are borne on separate plants with female organs located beneath lobed green caps elevated on stalks, like miniature umbrellas. They remind me of a group of commuters waiting in the rain for their bus.

LEAFY LIVERWORT COMMUNITY **Herbivores**
Small carrion beetle
Crane fly larva

LEAFY LIVERWORT

(*Lophocolea* sp.: Jungermanniales: Lophocoleaceae)

Characteristics: Moss-like trailing plant. Leaves (perianth) opposite, lacking central midrib. Capsules (spore cases) black; borne on tips of stalks (setae).

Regional habitat: Widespread in dune forests from Washington to California.

Comments: Together with mosses and lichens, liverworts form a living cover on the ground and over dead and living tree limbs in moist forests. Small carrion beetle larvae (Coleoptera: Leptodiridae) (left) feed on spores in the capsules. Crane fly larvae (*Limonia* sp.: Diptera: Tipulidae) (right) feed on liverworts growing on tree limbs.

LEAFY LIVERWORT
(*Radula* sp.: Jungermanniales: Radulaceae)
Characteristics: Leaves flattened, dark green; separate male and female sex organs. Thalli bear tiny brown spherical structures (gemmae) for asexual reproduction.

Regional habitat: Widespread in dune forests from Washington to California.
Comments: Some foreign representatives of this genus are known to contain cannabinoids.

THALLOID LIVERWORT

(*Marchantia* sp.: Marchantiales: Marchantiaceae)

Characteristics: Thalli flattened, dark green; female plants produce umbrella-like receptacles that contain reproductive structures.

Regional habitat: Widespread in dune forests from Washington to California.

Comments: Female plants of *Marchantia* resemble miniature umbrellas.

BIRDS-FOOT TREFOIL

Small clusters of yellow flowers and five-parted leaves, two of which are greatly expanded stipules, readily identify one of the standard pasture legumes, birds-foot trefoil. A perennial introduced from Europe as a forage crop, it tolerates poorly drained and acidic soils. The plant now has spread throughout most of the United States, growing in a variety of habitats. It is considered a valuable food for Canada goose, deer, and elk. The plants form nodules with *Rhizobium* bacteria that allow them to fix atmospheric nitrogen for faster growth.

Rosy trefoil, with small pinkish flowers, is a native perennial less common than birds-foot trefoil. Rosy trefoil is restricted to forests, stream banks, and wetlands along the Pacific coast and was never considered a pasture plant. Its long pods contain speckled pea-shaped seeds.

BIRDS-FOOT TREFOIL
(*Lotus corniculatus*: Fabaceae)
Characteristics: Low spreading, clover-like herb. Leaves divided into three to five leaflets. Flowers pea-like. large, yellow, often with reddish hues.
Regional habitat: Clearings in dune forests from Washington to California.
Comments: Bumblebees frequent the flowers and caterpillars and weevils devour the leaves. Birds-foot trefoil was introduced from Europe and is used as a fodder plant. It also occurs in deflation plains behind the fore dunes. Another trefoil with pinkish flowers and compound leaflets that is native to Western North America and found along the edges of dune forests, is rosy trefoil (*Hosackia rosea*) (right).

WESTERN SKUNK CABBAGE

Western skunk cabbage is not only one of the first flowers to bloom in the spring but is one of the strangest plants in dune forests. Numerous little flowers tightly arranged on a spadix look like an old-fashioned, elongate flashbulb with the large partially enclosing bract serving as a reflector. Also characteristic is the absence of a stem, with leaves emerging directly from the rootstalk. Small rove beetles can be found swarming over the flowers, stopping to feed on pollen and probably serving as pollinators. A tincture from the rootstalk provides a tea that has been used to treat cramps and control fits of coughing.

SKUNK CABBAGE COMMUNITY Herbivore
Rove beetle

WESTERN SKUNK CABBAGE
(*Lysichiton americanus*: Araceae)
Characteristics: Medium-sized perennial ground plant. Leaves large, broad, shiny, basal. Flowers numerous, small, bisexual; greenish with yellow anthers and stigmas; tightly appressed in stout spike (spadix) subtended by huge yellowish-green bract (spathe). Fruit berry-like, embedded in flower spike.

Regional habitat: Moist areas in dune forests from Washington to California.
Comments: Flowers appear before leaves in spring. The plant has an odor resembling a skunk, thus explaining one of its common names. It also is called swamp lantern.

ROVE BEETLE

(*Pelecomalium* sp.: Coleoptera: Staphylinidae)

Characteristics: Adult beetle small, dark brown with long antennae, narrow thorax and short, ocher-colored elytra.

Regional habitat: Dune forests from Washington to California.

Comments: These beetles are obligate pollen feeders and probably pollinate skunk cabbage flowers. Their larval biology is unknown.

PURPLE LOOSESTRIFE

Often reaching two meters in height, purple loosestrife, with its pinkish flowers, towers above most other herbaceous perennials. The species was introduced from Europe and very few American insects show much interest in it. That is why a little leaf beetle known to favor purple loosestrife was brought from Europe and released in the United States to try and control its spread. The black-spotted yellow larvae and brown-backed adults are commonly found feeding on leaves, but whether they can keep purple loosestrife under control before it takes over more forested and wetland areas waits to be seen.

PURPLE LOOSESTRIFE COMMUNITY **Herbivore**
Leaf beetle

PURPLE LOOSESTRIFE
(*Lythrum salicaria*: Lythraceae)
Characteristics: Tall erect perennial. Leaves simple, sessile, opposite to alternate. Flowers sessile, petals crinkled, pinkish-red; borne in racemes at stem tips.
Regional habitat: Borders of dune forests from Oregon to California.
Comments: Purple loosestrife is widespread and considered a noxious weed.

LEAF BEETLE

(*Galerucella calmariensis*: Coleoptera: Chrysomelidae)

Characteristics: Larva yellow with rows of dark bars and spots; head black. Adult dark brown; pronotum with central black spot; elytra with light border along outer margins.

Regional habitat: Wherever purple loosestrife occurs throughout the Pacific Northwest.

Comments: This leaf beetle was introduced into Oregon from Germany in 1992 as a biological control agent to reduce populations of purple loosestrife.

OREGON GRAPE

Yellow flower heads of Oregon grape are quite showy, as if flaunting their status as State Flower of Oregon. And the plant has been here for some time since fossil leaf impressions have been found on rocks along the coast. Oregon grape produces the alkaloid, berberine, and its roots have been used to treat microbial infections on skin and in the alimentary tract. The leaves are so tough that it is surprising to find two different caterpillars able to digest them. The insects' microhabitats are different, however. Leaf caterpillars chew through the entire leaf while leaf miners feed within the leaflets and use the tough epidermal layers for protection.

OREGON GRAPE COMMUNITY Herbivores	 Parasite
Leaf caterpillar	Wasps
Leaf miner	

OREGON GRAPE

(*Mahonia aquifolium*: Berberidaceae)

Characteristics: Erect evergreen shrub. Leaves alternate; leaflets opposite, stiff, holly-like with spiny margins. Flowers bright yellow; borne in clusters at stem tips. Fruits are blue berries.

Regional habitat: Edges of dune forests from Washington to California.

Comments: Berries are quite tart but have been used for jelly and are said to have medicinal properties.

LEAF CATERPILLAR

(Lepidoptera: Ypsolophidae)

Characteristics: Caterpillar brownish-gray with longitudinal white lines and a series of "eyespots" running length of body; posterior pair of pseudolegs usually protruding from body tip. Cocoon very durable, constructed within folded leaf.

Regional habitat: Dune forests in Oregon.

Comments: Adult moths were not obtained, mainly because all caterpillars collected were parasitized by wasps (Hymenoptera: Ichneumonidae) (bottom) and never pupated.

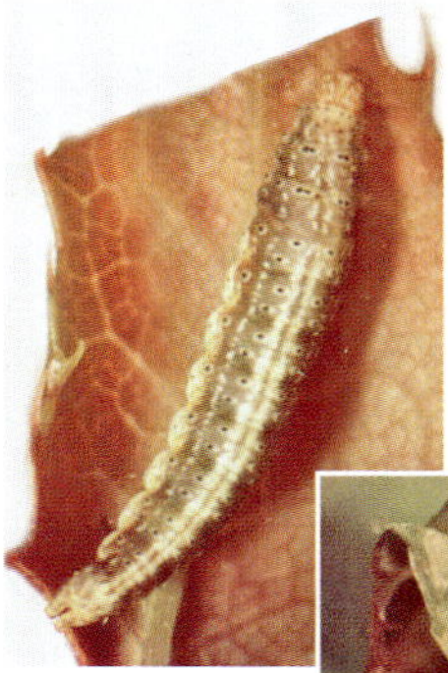

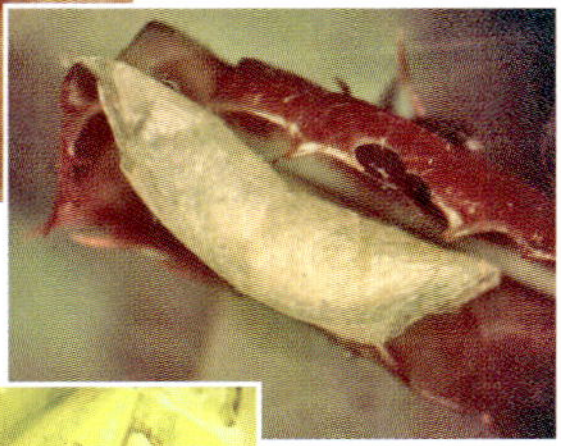

LEAF MINER

(*Bucculatrix* sp.: Lepidoptera: Bucculatricidae)

Characteristics: Caterpillar small, yellow, flattened; tapering toward tail end, head with brown markings; legs reduced.

Regional habitat: Dune forests in Oregon.

Comments: The "mine" is a wavy band usually starting at the margin of a leaflet and then moving inwards as the caterpillar grows. All development occurs inside the mine; the caterpillar is protected by the leaflet's stiff epidermal cells.

CALIFORNIA BAYBERRY

You can smell the spicy odor emanating from the sticky leaves and fruits of bayberry even before you see their evergreen sprigs. But the chemicals producing the scent can be quite allergenic, which I learned after letting a friend smell a leaf I had just crushed. The fumes closed his nasal passages and he was gasping for breath immediately after taking the first whiff. Fortunately, the swelling passed but it taught me something about this native plant.

The same allergenic compounds may be able to repel insect attacks, but some bugs have managed to circumvent them. Along the west coast occurs a leaf roller moth whose caterpillars are able to develop on the pungent leaves of California bayberry. While light green caterpillars that feed in the folded leaves all appear to be identical, the adult stages have quite variable color patterns. Whether these differences represent separate species or color variants of the same species (polymorphs) is unknown.

It always amazes me how a little fly can provoke a large plant gall that can support the development of all its progeny. In California bayberry, gall gnat maggots induce swollen reddish stem galls that resemble a human skin reaction to a bee or wasp sting. Pupae wiggle partially out of the gall tissue, allowing the long-legged, flimsy adults an easy escape route.

CALIFORNIA BAYBERRY COMMUNITY Herbivores	Parasites
Varied leaf roller moth	Tachinid flies
Gall gnat	Wasps
Omnivorous looper	
Black-backed stink bug	

CALIFORNIA BAYBERRY

(*Myrica* (*Morella*) *californica*: Myricaceae)

Characteristics: Shrub or small tree. Leaves shiny, evergreen, leathery, scented; glandular yellow dots on undersurface. Minute male and female flowers borne in separate clusters along branches (monoecious). Ripe fruit a reddish-brown, waxy, nutlike berry.

Regional habitat: In and adjacent to moist dune forests from Washington to California.

Comments: The reddish fruits are somewhat waxy, but are eaten by birds.

VARIED LEAF ROLLER MOTH

(*Epinotia emarginana*: Lepidoptera: Tortricidae)

Characteristics: Caterpillar light green; head tan; develops inside rolled up leaves. Pupa light reddish brown; formed within caterpillar feeding chamber. Adult coloration extremely variable.

Regional habitat: Dune forests in Oregon and California.

Comments: The great range of colors on adult moths **(B)** indicates a high rate of genetic re-shuffling and suggests that this is an evolving species. Caterpillars are parasitized by tachinid flies (Diptera: Tachinidae) and ichneumon wasps (Hymenoptera: Ichneumonidae). The tachinid fly maggot forms a black puparium adjacent to the brown pupa of the leaf roller **(A)**. The emerging fly is dark and robust with clear wings **(C)**. The ichneumonid wasp pupa is a white fuzzy cocoon that is formed adjacent to the shriveled host **(D)**.

GALL GNAT

(Diptera: Cecidomyiidae)

Characteristics: Maggot small, yellowish; develops inside reddish globular galls at tips of bayberry branches. Mature pupa white, flexible; wiggles partially out of gall before adult emergence. Empty pupal case remains attached to gall. Adult gnat small with long legs.

Regional habitat: Dune forests in Oregon.

Comments: I could not find a record of this gall gnat on California bayberry.

OMNIVOROUS LOOPER

(*Sabulodes aegrotata*: Lepidoptera: Geometridae)

Characteristics: Caterpillar with greenish back and two wide black stripes along sides of body; two pairs pseudolegs. Adult dark reddish-brown; fore- and hindwings with faint dark bands.

Regional habitat: Dune forests in Oregon and California.

Comments: Caterpillars are quite variable in color patterns and often make webbing around themselves while feeding.

RED-BACKED STINK BUG

(*Banasa dimiata*: Hemiptera: Pentatomidae)

Characteristics: Adult medium-sized, greenish; base of thorax with red to rusty bar; wing covers with red to rusty patches.

Regional habitat: Widespread throughout the Pacific Northwest.

Comments: These bugs are herbivorous and use their long, stout beaks to imbibe plant juices. They feed on fruits of California bayberry.

SITKA SPRUCE

How strange to see the remains of a forest in the ocean. Between Island Hill at Neskowin, Oregon and the outgoing tide are dozens of Sitka spruce stumps, evidence that, some 2,000 years ago, a healthy forest flourished where there is now salt water. Apparently, the subsidence of this Sitka spruce forest was caused by an earthquake following a volcanic eruption. The forest must have been fairly extensive since the protruding stumps are visible for quite some distance. As the outgoing salt water tide swirled around the trunks, I thought how amazing it was that they've lasted as long as they have. Many stumps were covered with myriads of small white barnacles.

One can imagine the events that occurred on that horrific day when the earth rumbled, the ground split open, the embankments collapsed, and an entire forest crashed down into the sea. Many organisms lost their lives as the waves thundered up and began engulfing the tree trunks. Nothing in their genetic makeup prepared them for such an event. Racing up to the top of the immersing spruce trees was the only possible route to survival, but even that was of little avail, since eventually all the trees were engulfed in water.

All terrestrial plants have some insects feeding on them and conifers are no exception. Conifers may have evolved resin production to repel pests, but insects are infinitely adaptable. Thus, it is possible to find herbivorous caterpillars, weevils, bark beetles, and longhorn beetles nestled among the needles or within the bark of conifers such as Sitka spruce. Small eruptant rust fungi emerging from the needles are just one of many plant pathogens found on conifers.

SITKA SPRUCE COMMUNITY			
Herbivores	**Parasites**	**Pathogens**	**Commensals**
Budworm	Flies	Needle rust	Nematodes
Root weevil	Wasps	Burl	Mites
Engraver beetle			
Yellow-headed sawfly			
Silver-spotted tiger moth			
Longhorn beetle			

SITKA SPRUCE

(*Picea sitchensis*: Pinaceae)

Characteristics: Large evergreen conifer. Needles short, stiff, flattened, sharply pointed; borne singly on pegs with two white lines on both surfaces. Seed cones with thin, wavy, overlapping scales lacking prickles.

Regional habitat: Dune forests from Washington to California.

Comments: Sitka spruce is long-lived, with a lifespan of several centuries. Some 200 Sitka spruce stumps in the sea form a ghost forest in Neskowin, Oregon. These stumps, which are only exposed at very low tides, were part of a forest that existed some 2000 years ago along the Pacific coast. Chemicals in the salt water preserved these stumps to the present day.

BUDWORM

(*Choristoneura* sp.: Lepidoptera: Tortricidae)

Characteristics: Caterpillar green; prothorax with black band; head black to dark brown. Pupa dark brown; formed in feeding site of caterpillar. Adult silvery-brown; forewings with a few darker brown markings.

Regional habitat: Dune forests from Washington to California.

Comments: Caterpillars bind needles at branch tips to form a protective domicile where they feed and pupate. They are attacked by parasitic wasps.

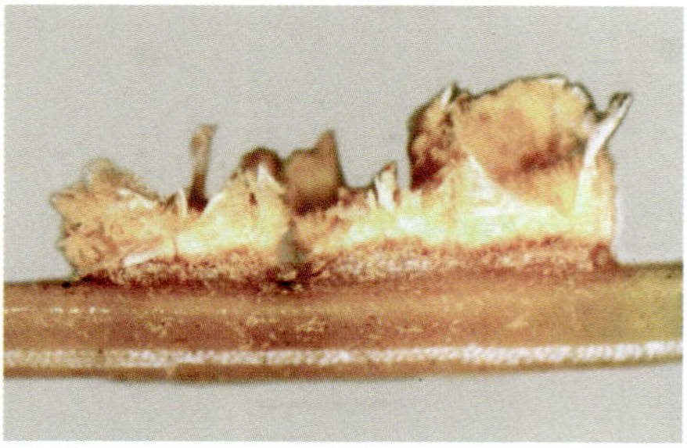

NEEDLE RUST

(*Chrysomyxa ledicola*: Pucciniales: Coleosporiaceae)

Characteristics: Pale orange spore masses emerge in brown "dead" zones on spruce needles.

Regional habitat: Dune forests from Washington to California.

Comments: Labrador tea (*Rhododendron groenlandicum*) is the alternate host and is required to complete the life cycle of spruce needle rust.

ROOT WEEVIL

(*Sthereus quadrituberculatus*: Coleoptera: Curculionidae)

Characteristics: Adult small, brown; beak long; four tubercles on fused elytra; each tubercle bears a group of erect black hairs.

Regional habitat: Dune forests from Washington to California.

Comments: Larval stages probably feed on rootlets of Sitka spruce. Flightless adults nibble on needles and rest in ground litter.

ENGRAVER BEETLE

(*Dryocoetes caryi*: Coleoptera: Scolytidae)

Characteristics: Larva small, white; legless; develops in wood. Adult slender, reddish-brown, head hidden from above, elytra covered with longitudinal rows of pits.

Regional habitat: Dune forests from Washington to California.

Comments: Mites and nematodes are often present under the elytra.

BURL

Characteristics: Large bulbous woody growth, up to a meter in diameter, on Sitka spruce trunk.

Regional habitat: Only found on Sitka spruce in dune forests in Washington.

Comments: Although the origin of these growths is unknown, the cause is most likely an infectious agent. Perhaps bark beetles that develop within the wood introduce some pathogen that initiates the burl.

YELLOW-HEADED SAWFLY

(*Pikonema* sp.: Hymenoptera: Tenthredinidae)

Characteristics: Larva green with several black stripes running length of body; head yellow to reddish-brown depending on geographic location. Pupa forms in reddish-brown fibrous cocoon. Adult small, reddish-brown, wasp-like.

Regional habitat: Dune forests from Washington to California.

Comments: When abundant, these larvae can defoliate entire trees. Cocoons are formed in debris or soil under the tree. Parasitic wasps and flies attack the larvae but some may be repelled by yellow defensive glands that are extruded from the neck of the larva.

SILVER-SPOTTED TIGER MOTH

(*Lophocampa argentata*: Lepidoptera: Arctiidae)

Characteristics: Early stage caterpillars are yellow with black spots, mass together in common web. Older caterpillar darker, covered with reddish hairs; alternating yellow and black areas. Pupa reddish-brown. Cocoons loose. Adult brownish; forewings with a series of white spots.

Regional habitat: Dune forests from Washington to California.

Comments: Caterpillars often collect together in a common web when young, then become solitary as they mature. Caterpillars are attacked by parasitic wasps and flies. Often the fly parasites are carried into the moth's pupal stage and form a puparium that protrudes from the anus of the moth's pupa.

LONGHORN BEETLE

(*Judolia* sp.: Coleoptera: Cerambycidae)

Characteristics: Larva white; legless; develops in Sitka spruce wood. Adult large; head and thorax dark; elytra light with large dark brown blotches.

Regional habitat: Dune forests from Washington to California.

Comments: While the larvae feed inside the trunks and branches of Sitka spruce, the adult beetles visit flowers of various plants, such as sea coast angelica, salmonberry, and Nootka rose for nectar and pollen. Mites and nematodes can be found under the beetle's wing covers.

DOUGLAS-FIR

Douglas-fir is a native western North America conifer well adapted to moist, cool dune forests. It ranges along the Pacific Northwest coast from British Columbia to central California. These trees can live up to 1000 years, which explains why they are often so tall. Douglas-fir has permanent connections to two Scotsmen, the botanist, David Douglas, who provided its common name and the physician, Archibald Menzies, who provided the species name.

Along the coast, the trees can extend over 100 meters in height. Foliage is so dense that voles live among the boughs. Maybe that is why spotted owls are often seen resting on the limbs of Douglas-fir. The sap is quite resinous and difficult to remove from skin and clothes. Coastal peoples used it for patching holes in their dwellings and boats. It was also used as a treatment for wounds since it contains antibiotic properties and might protect a wound from foreign microbes. Today the wood of Douglas-fir is widely used for construction and its rapid growth makes it a suitable commercial, ornamental "Christmas" tree.

Douglas-fir has formed a symbiotic association with an ectomycorrhizal fungus that forms slippery jack mushrooms as its perfect (producing sexual spores) stage. The mushrooms don't last long once discovered by banana slugs. A primitive insect that feeds on Douglas-fir logs on the forest floor is the wood roach. Feeding on wood is a challenge since most insects can't break down the fibers. However, the wood roach receives assistance from colonies of flagellated protozoa living inside its gut, very similar to those found in termites.

Some organisms have become dependent on fallen Douglas-fir cones. The cone mushroom competes with cone caterpillars and cone borer beetles for this resource.

DOUGLAS-FIR COMMUNITY		
Herbivores	Detritivores	Parasite
Cone moth	Conifer cone mushroom	Wasps
Needle binder	Blue-staining slippery jack	
Green zebra moth	Wood roach	
Cone borer beetle		
Longhorn beetle		
Needle scale		
Banana slug		
Green flathead borer		
Conifer seed bug		

COAST DOUGLAS-FIR

(*Pseudotsuga menziesii* var. *menziesii*: Pinaceae)

Characteristics: Large erect conifer. Needles flat, with pointed tips, spirally arranged, with two white stomatal lines on lower surface. Pollen (male) cones small, reddish, mostly erect, clustered at branch tips. Seed (female) cones with three-forked bracts (center one usually narrow) extending beyond cone surface, hanging from branch tips.

Regional habitat: Dune forests from Washington to central California.

Comments: This is the second tallest conifer in the world, after the giant redwood. Its cones can be identified by long bracts that extend over the seed scales and provide a runway for rainwater. The needles are bound together by caterpillars for makeshift shelters.

CONIFER CONE MUSHROOM

(*Strobilurus trullisatus*: Agaricales: Physalacriaceae)

Characteristics: Delicate, small, white to pinkish or brownish mushroom with a slender stalk, dry cap and white gills and spores.

Regional habitat: Dune forests from Washington to California.

Comments: This little mushroom occurs only on fallen cones, in particular those of Douglas-fir. Snails devour the caps.

NEEDLE SCALE

(*Chionaspis pinifoliae*: Hemiptera: Diaspididae)

Characteristics: Scales small, yellowish to brownish; oyster-shaped; arranged in rows on needles.

Regional habitat: Dune forests from Washington to California.

Comments: When populations of scale insects are high, some yellowing of needles and even dieback of branches can occur. Small wasps parasitize the scales.

CONE MOTH

(*Dioryctria* sp.: Lepidoptera: Pyralidae)

Characteristics: Caterpillar reddish-brown; thorax and terminal segment brown; head dark brown. Pupa brown; formed inside cone. Adult small, brownish-gray; forewings with white triangular patch near tips.

Regional habitat: Dune forests from Washington to California.

Comments: Several different moths develop in Douglas-fir cones. Members of this genus (*Dioryctria*) are widespread in North America.

GREEN ZEBRA MOTH

(*Feralia deceptiva*: Lepidoptera: Noctuidae)

Characteristics: Caterpillar green; series of white bands running length of body. Adult moth with gray-green forewings crossed with irregular black and white lines.

Regional habitat: Dune forests from Washington to California.

Comments: Caterpillars feed on needles of Douglas-fir. Adults are nocturnal.

CONE BORER BEETLE

(*Ernobius punctulatus*: Coleoptera: Ptinidae)

Characteristics: Larva small, white; head capsule tan; completely embedded within female cone. Adult beetle small, elongate with dark thorax and head and reddish-brown elytra.

Regional habitat: Dune forests in Washington and Oregon.

Comments: This species has only been reported from Douglas-fir. Related species occur in cones of other conifers.

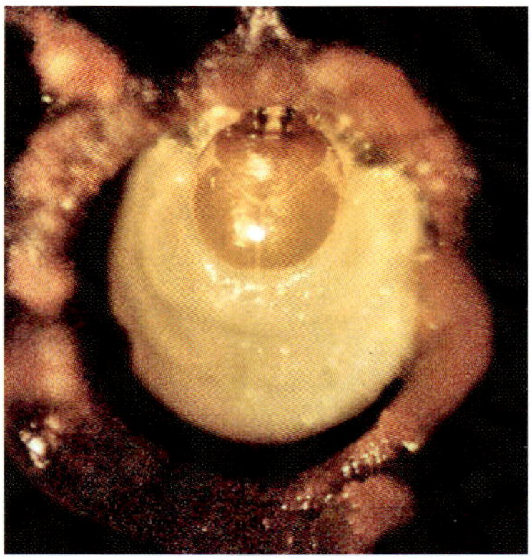

LONGHORN BEETLE

(*Prionus californicus*: Coleoptera: Cerambycidae)

Characteristics: Developing larva white; body segments swollen; head dark brown; legs minute. Body darkens soon after death, as depicted. Adult beetle dark reddish-brown; pronotum bearing three marginal spines; antennae long, thick.

Regional habitat: Dune forests from Washington to California.

Comments: This is one of the largest beetles in the Pacific Northwest. The larvae may feed for two to three years on roots of Douglas-fir before pupating.

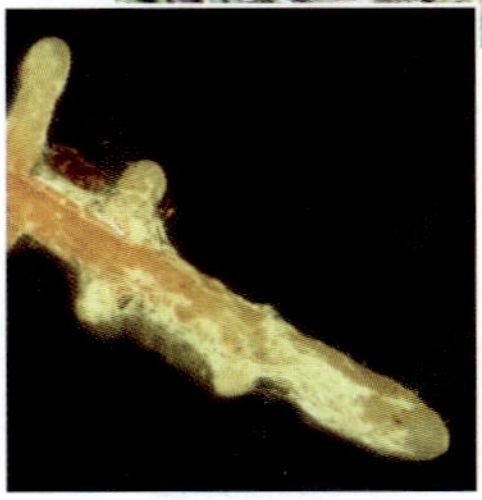

BLUE-STAINING SLIPPERY JACK

(*Suillus tomentosus*: Boletales: Suillaceae)

Characteristics: Large mushroom. Stalk thick, yellow. Cap orange-yellow, scaly; spores brown. Flesh turns blue when bruised.

Regional habitat: Throughout the Pacific Northwest.

Comments: These mushrooms are a favorite food for banana slugs (*Ariolimax columbianus:* Gastropoda: Ariolimacidae). Slippery jack is probably the perfect (producing sexual spores) stage of an ectomycorrhizal fungus shown here covering a rootlet of Douglas-fir.

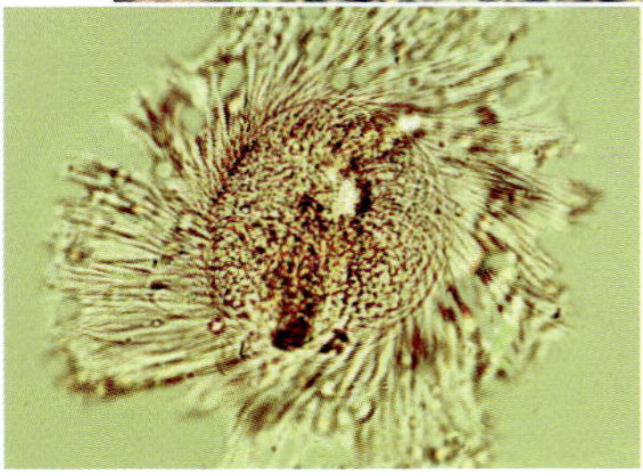

WOOD ROACH

(*Cryptocercus punctulatus*: Blattodea: Cryptocercidae)

Characteristics: Newly hatched roach whitish, nearly transparent; antennae extended. Mature adult reddish-brown to black, shiny, metallic; antennae concealed.

Regional habitat: Dune forests from Washington to California.

Comments: Wood roaches are wingless and feed on rotting logs of Douglas-fir and other trees. They are sub-social (live in family groups) and are similar to the lower termites in harboring mutualistic flagellated protozoa in their guts to digest wood **(bottom)**.

GREEN FLATHEAD BORER

(*Buprestis langi*: Coleoptera: Buprestidae)

Characteristics: Adult large, metallic green, oval, flattened; front of head rounded; antennae short, slender. Females completely green; males with some dark markings on elytra.

Regional habitat: Forests from Washington to northern California.

Comments: The large white larvae develop in Douglas-fir wood.

NEEDLE BINDER

(*Acleris* sp.: Lepidoptera: Tortricidae)

Characteristics: Caterpillar yellowish; head and thorax dark brown.

Regional habitat: Dune forests throughout the Pacific Northwest.

Comments: Caterpillars web together adjacent needles and feed within the shelter.

CONIFER SEED BUG

(*Leptoglossus occidentalis*: Hemiptera: Coreidae)

Characteristics: Large, elongate, mostly dark brown bug with a pointed head, protruding compound eyes, fairly long antennae, finely speckled forewings with many veined membrane. The expanded hind tarsi are characteristic of this group, which are known as leaf-footed bugs.

Regional habitat: Dune forests from Washington to California.

Comments: This species feeds on the sap of developing seeds of Douglas-fir and other conifers, thus explaining its common name. Such feeding can damage the seeds.

FERNS

One of the pleasures of walking through dune forests is encountering the many types of ferns that beautify the landscape. Some dune forest ferns, like wood fern, deer fern, sword fern, and Pacific polypody, are evergreen, while others lose their leaves in autumn.

Ferns love moist, damp conditions, and are usually found with moss and liverworts, especially the leathery polypody fern, that grows on logs. This fern does not have the typical spreading leaves of most ferns. Leathery polypody not only has tough leaves, as its name implies, but it has developed an amazing tolerance to darkness. Very little sunlight reaches the dune forest floor and leathery polypody is usually the only fern growing on logs completely overshadowed by other vegetation.

When encountering deer fern, it first appears that there are two ferns growing together, one with spreading filmy leaves and the other with stiff erect fronds with short lobes. However, these represent separate sterile and fertile fronds of the same plant. In contrast, all leaves are similar in most other ferns. Deer fern produces spores on the upright fronds, which disappear after reproduction. Sterile fronds with filmy leaves persist and supply food for deer and elk throughout the winter. A story relates that antibiotic properties of deer fern were discovered when a deer was observed rubbing its antler stubs against the leaves.

Of all dune forest ferns, the wood fern is the most gracious, with spreading finely dissected fronds that at first are rolled up like the scroll of a violin. The little fern knotting gall fly seeks out these curled tips to lay its eggs. Some component in the eggs keeps the fern tips from unrolling, even making them roll tighter, which protects the developing maggots. This miniature "fern grotto" makes an ideal home and maggots remain feeding inside until pupation. Dark adult flies, with their large semi-circular heads, emerge from the galls and are soon off searching for mates.

Most ferns don't attract many insects; however, western sword fern is an exception. This tall evergreen fern with stiff erect leaves occurs all along the Pacific Northwest coast. Ferns are ancient plants and at least two species of sawflies, considered primitive hymenopterans, develop on western sword fern in dune forests, probably indicating an association lasting millennia. Other insects feeding on western sword fern are the broad-nosed weevil and several moth caterpillars. One of the latter is a small tortricid caterpillar that spins a silk safety net over itself to insure that it won't fall while feeding on the underside of the leaf. Larger, brown-lined noctuid caterpillars spin no safety nets, but curl up and tightly hold on to the leaf when disturbed. Most interesting are hover flies that, although normally going to flowers for food, can be seen probing spore sacs with their mouthparts. The spores probably serve as a nutrient source, just as pollen and nectar do on various flowers.

Spiders are everywhere and several types occur on sword ferns, including some orb-weavers that construct beautiful circular webs composed of multiple strands of silk.

Because sword ferns usually grow in humid habitats that favor fungal growth, it is not uncommon to find dead spiders covered with spores of parasitic fungi.

FERN COMMUNITY Herbivores	Parasites and predators
Two-spotted sawfly	Spider fungus
Green sawfly	Wasps
Short-nosed weevil	Orb-weaver spider
Leaf caterpillar	
Brown-lined caterpillar	
Hover fly	
Fern knotting gall fly	
Case bearer moth	
Aphids	

NORTHERN WOOD FERN
(*Dryopteris expansa*: Dryopteridaceae)
Characteristics: Medium-sized fern with scaly stalk. Fronds spreading, three-times pinnate with ultimate segments toothed; opposite pinnae usually unequal in size.

Regional habitat: Dune forests from Washington to California.
Comments: This is a delicate fern that only grows under damp, shady conditions. It is also known as the buckler fern.

FERN KNOTTING GALL FLY

(*Chirosia* sp.: Diptera: Anthomyiidae)

Characteristics: Maggot yellowish, legless, plump, with dark mouth hooks. Pupation occurs inside the fern gall. Adult blackish; eyes large; wings fuscous.

Regional habitat: Dune forests of Oregon and California.

Comments: Maggots develop inside loose galls formed by the coiling of the tips of northern wood fern fronds.

HOVER FLY

(*Eupeodes lapponicus*: Diptera: Syrphidae)

Characteristics: Maggot stout, yellowish-orange; legless (moves by undulating body). Puparium tan, egg-shaped, smooth with pair of small protruding terminal spiracles. Adult abdomen black, marked by transverse yellow bars; thorax bronze; eyes large, brown; wings transparent.

Regional habitat: Dune forests throughout the Pacific Northwest.

Comments: Maggots feed on aphids that infest northern wood fern. Puparia are attached to fern leaves. Adults feed on spores of northern wood fern as well as on pollen from nearby flowers.

DEER FERN

(*Blechnum spicant*: Athyriales: Blechnaceae)

Characteristics: Bushy evergreen fern with two frond types (sterile and fertile): 1) sterile fronds spreading with normal size leaflets; 2) fertile fronds upright with short, narrow, curved leaflets bearing two rows of spore sacs (sori).

Regional habitat: Dune forests from Washington to California.

Comments: Deer fern provides winter food for deer and elk. Deer rub their antler stubs on the plants, possibly obtaining some therapeutic effect.

CASE BEARER MOTH

(*Coleophora* sp.: Lepidoptera: Coleophoridae)

Characteristics: Caterpillar small; lives and pupates within self-made parchment-like case attached to leaflets. Pupa brownish; formed within case. Adult small, brownish, slender, with pointed head and erect antennae.

Regional habitat: Dune forests in Oregon.

Comments: Pupal cocoons are attached to fronds of deer fern.

LEATHERY POLYPODY
(*Polypodium scouleri*: Polypodiales: Polypodiaceae)
Characteristics: Small epiphytic evergreen fern with few short, thick, leathery leaves; leaflets broad, round-tipped with round spore balls (sori) on undersurfaces.
Regional habitat: Dune forests from Washington to California.
Comments: Leathery polypody survives quite well growing on tree branches, rocks, and fallen logs in heavily shaded areas under conifers. This gives the species an advantage over more aggressive plants that require additional light for survival.

PACIFIC POLYPODY
(*Polypodium amorphum*: Polypodiales: Polypodiaceae)

Characteristics: Evergreen. Fertile and sterile fronds similar. Leaf blades divided once to form wide, entire, rounded tips.

Regional habitat: Edges and open areas of dune forests from Washington to California.

Comments: This species requires more light than leathery polypody. Rhizomes were eaten and used for medical purposes by Native Americans.

WESTERN SWORD FERN
(*Polystichum munitum*: Dryopteridaceae)
Characteristics: Tall evergreen fern with erect, stiff, once pinnate leaves; leaflets alternate, pointed, spinulose-toothed with spores groups (sori) in rows on under surface.

Regional habitat: Dune forests and coastal dunes from Washington to California.
Comments: Sword fern attracts a variety of insects.

TWO-SPOTTED SAWFLY

(*Strongylogaster* sp.: Hymenoptera: Tenthredinidae)

Characteristics: Larva yellowish; head tan with black eye spots. Cocoon oval, brown with rough surface. Adult yellowish-green; head and thorax with black spots; eyes and antennae black.

Regional habitat: Dune forests from Washington to California.

Comments: The light color of adults along with a black head and thoracic spots makes them easy to recognize.

GREEN SAWFLY

(*Aneugmenus* sp.: Hymenoptera: Tenthredinidae)

Characteristics: Larva green, elongate; head tan. Adult brown to black; legs tan.

Regional habitat: Dune forests from Washington to California.

Comments: Larvae eat spore sacs as well as leaves. Two different parasitic ichneumonid wasps (Hymenoptera: Ichneumonidae), one brown and one green, attack the larvae.

SHORT-NOSED WEEVIL

(*Sciopithes* sp.: Coleoptera: Curculionidae)

Characteristics: Adult mottled brown; eyes dark brown, conspicuous; beak short; antennae long.

Regional habitat: Dune forests from Washington to California.

Comments: Adults feed on fern leaves while larvae develop on roots of various plants.

LEAF CATERPILLAR

(*Acleris* sp.: Lepidoptera: Tortricidae)

Characteristics: Caterpillar reddish-tan; head black.

Regional habitat: Dune forests in Washington and Oregon.

Comments: Caterpillars spin silken webs around spore masses (sori) and feed on the spores.

BROWN-LINED CATERPILLAR

(*Phlogophora* sp.: Lepidoptera: Noctuidae)

Characteristics: Eggs white, oval; deposited in irregular rows on fern leaflets. Caterpillar light brown with a series of short diagonal, brown dorsal stripes.

Regional habitat: Dune forests in Oregon.

Comments: Eggs are deposited on the underside of leaves. Caterpillars move from leaf to leaf and coil when disturbed.

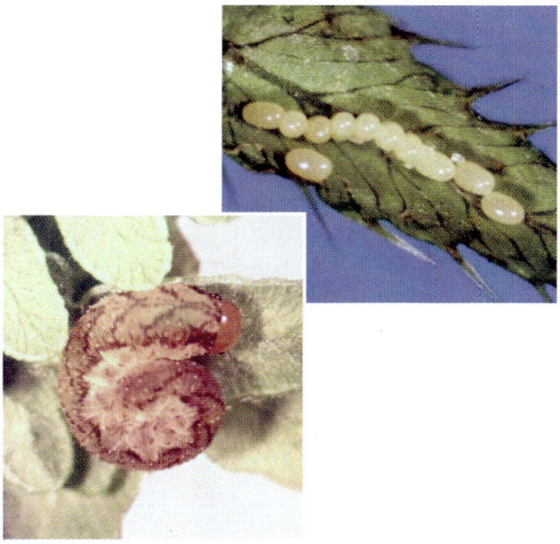

ORB-WEAVER SPIDER

(*Nuctenea* sp.: Araneae: Araneidae)

Characteristics: Small spider with rounded abdomen of various patterns; eight eyes; spiny legs.

Regional habitat: Dune forests throughout the Pacific Northwest.

Comments: Orb-weavers spin circular webs stretching between leaves of sword fern. They are attacked by *Gibellula* (Cordycipitaceae) fungi that sporulate on the dead spider's body.

CASCARA

The shiny, oval, ribbed leaves of cascara are much more obvious than the tiny flower clusters emerging from the leaf bases. Mature purple to black cherry-like fruits that can be eaten raw or cooked are still more evident. However, it is the bark that has made cascara famous. A tincture made from the bitter bark of the limbs provides an effective laxative. In fact, it works so well that commercial preparations were formulated and sold globally. Since cascara is native only to mixed deciduous-conifer forests of the Pacific Northwest, populations declined as plants were removed for their bark. Apparently, the bitter compounds are not carried over into cascara honey, although even that product is said to be slightly laxative in effect. It is curious what other uses have been made of the bitter bark, such as applying a tincture to fingernails to stop nail biting. A related plant with similar properties is *Rhamnus cathartica,* or common buckthorn, a spiny branched native of Europe, northwest Africa, and western Asia. Chemicals in seeds and leaves of that plant also have a laxative effect that, at too high a dose, can cause stomach cramps. Common buckthorn was introduced into North America and is considered an invasive species is some areas.

CASCARA COMMUNITY **Herbivores**
Looper
Leaf gall moth
Leaf roller
Leaf mining fly

CASCARA

(*Rhamnus purshiana*: Rhamnaceae)

Characteristics: Large deciduous shrub. Leaves simple, alternate, oval, veins parallel, margins minutely toothed. Flowers small with five whitish petals borne in clusters at tips of short stems (peduncles) arising from leaf axils. Mature fruit a blackish-blue drupe.

Regional habitat: Dune forests from Washington to California.

Comments: Cascara, a native of the Pacific Northwest, is also known as buckthorn, but should not be confused with the exotic common buckthorn, *Rhamnus cathartica*.

LOOPER

(*Triphosa haesitata*: Lepidoptera: Geometridae)

Characteristics: Caterpillar light green; faint white longitudinal lines running length of body. Pupa mostly green with yellowish abdomen. Adult large with short antennae; forewings with many light to dark brown, transverse wavy lines.

Regional habitat: Dune forests from Washington to California.

Comments: It is unusual for a moth to have a green-colored pupa.

LEAF GALL MOTH

(*Sorhagenia nimbosus*: Lepidoptera: Cosmopterigidae)

Characteristics: Caterpillar small, plump, white with short brown pronotum and black head. Adult small, narrow, grayish-brown; expanded forewing tips with dark spots.

Regional habitat: Dune forests in Oregon.

Comments: Caterpillars are quite rotund and live completely inside the leaf gall.

LEAF ROLLER

(*Acleris* sp.: Lepidoptera: Tortricidae)

Characteristics: Caterpillar small, gray; pronotum with black band; head black. Pupa brown. Adult mostly white; each forewing with two bold, brown median crescents.

Regional habitat: Dune forests in Oregon.

Comments: Caterpillars feed and pupate within rolled up leaves.

LEAF MINING FLY

(*Phytomyza* sp.: Diptera: Agromyzidae)

Characteristics: Maggot small, green, with dark mouth hooks. Puparium small, white; walls partially transparent. Adult fly small, gray; eyes reddish; wings clear.

Regional habitat: Dune forests in Washington and California.

Comments: Maggots make sinuous tunnels as they eat their way through tissue between upper and lower leaf surfaces. The pupal case is partially transparent, which makes it possible to follow the formation of the adult fly.

SNAKEBERRY

The large, shiny leaves of snakeberry add a deep green color to the forest floor. Grooves on heart-shaped leaves partially surrounding the flower spikes function as spouts that lead rivulets of moisture to the ground. Snakeberry plants might be called social since they usually grow so close to each other that the forest floor beneath them is hidden from view. Early in the summer, a spike bearing delicate white flowers appears like a candle with a long white flame.

Snakeberry often grows in dense shade. Little yellow sawfly larvae feed on snakeberry, hiding during the day and emerging at night to chew large holes in the leaves. By late summer, large portions of the attacked leaves are missing. Where the pupae are formed is still a mystery.

In a dune forest in California, I came across a curious insect larva feeding on snakeberry leaves. It was so strange that I couldn't even tell its head from its tail. I mused that this creature must have come from outer space since it had no legs and a strange pointed head. The pupa was even more bizarre. It looked like a tube with a pair of protruding devil's horns. The tail portion was enclosed in the transparent skin of the last larval stage, which was glued to the leaf surface. Obviously, just before it pupated, the larva only shed a portion of its skin, leaving the remainder attached to the leaf. I had never seen such behavior previously. How it was able to secure its skin to the leaf surface is a mystery. From the long slender pupa eventually emerged a gangly, long-legged adult cranefly. This was a surprise since crane fly larvae normally occur in the soil and don't feed completely exposed on leaves.

The larva of another fly that develops on snakeberry is a leaf miner. It forms brown puparia inside the leaf and emerges as a delicate, long-legged, brown fly.

SNAKEBERRY COMMUNITY Herbivores	Parasite
Sawfly	Wasp
Crane fly	
Leaf miner	

SNAKEBERRY

(*Maianthemum dilatatum*: Asparagaceae)

Characteristics: Small deciduous perennial herb with erect stems. Each plant usually has only two or three, broad, shiny, heart-shaped leaves with parallel veins. Flowers small, white; borne in elongate terminal clusters. Mature berries reddish.

Regional habitat: Dune forests from Washington to California.

Comments: Snakeberry provides a dominant ground cover, especially in Sitka spruce forests. It has been used to treat medical conditions ranging from sore eyes to stomach pains. Other names of the plant are false lily of the valley and May lily.

SAWFLY

(*Phymatocera* sp.: Hymenoptera: Tenthredinidae)

Characteristics: Larva yellow, covered with tiny black spots; head black. Adult black; head wide; antennae outstretched; wings black-tinted.

Regional habitat: Dune forests from Washington to California.

Comments: Larvae are nocturnal and make large roundish holes in the leaves as they feed. Adults are very difficult to find and the location of pupae is unknown.

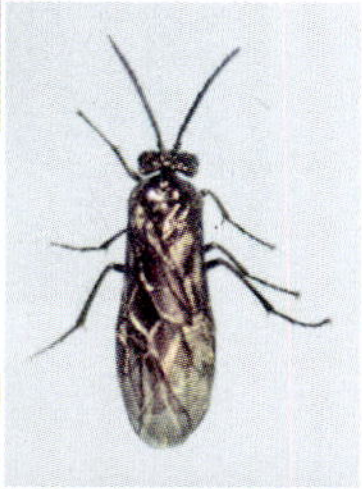

CRANE FLY

(*Cylindrotoma distinctissima americana*: Diptera: Cylindrotomidae)

Characteristics: Larva green, legless. Body wall transparent, revealing gut filled with brown residue. Pupa green with two short head projections; attached to leaf by brown, partially shed, last larval instar cuticle. Adult brown, spindly; head small; abdomen long; legs extended; wings narrow.

Regional habitat: Dune forests in California.

Comments: This is one of the few species of crane flies that feeds exposed on the leaves of living plants.

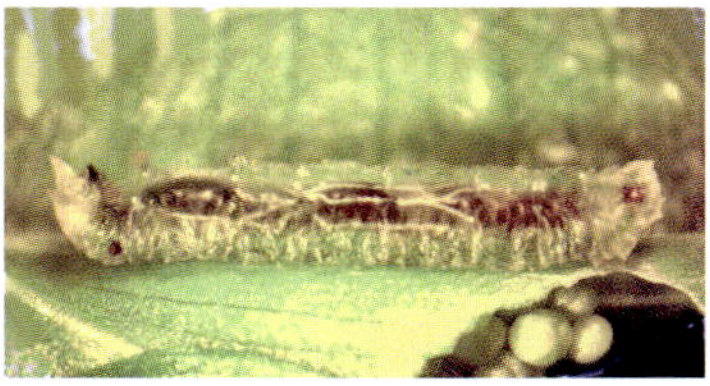

LEAF MINER

(*Parallelomma* sp.: Diptera: Scathophagidae)

Characteristics: Maggot white, elongate, with black mouth hooks. Puparium brown with pair of projections at posterior end. Adult delicate, slender, brown with large reddish eyes, long legs and short antennae; wings transparent, with rounded tips.

Regional habitat: Dune forests in Oregon.

Comments: Larvae, which are attacked by parasitic wasps, form blotch mines in snakeberry leaves.

MOSSES

Mosses provide carpets for the forest floor. They cover rocks, fallen stumps and the soil surface with a variety of colors and textures. Some are silky, almost velvety, with flattened leaves and extended stems. Running your hand over them is like petting a fluffy cat. Others are coarse and somewhat prickly with erect stems and spiny green leaves. Every moss has its own characteristics. While some barely show themselves and live in the shadows, others are eminently noticeable, struggling to compete for the sparse sunlight that passes through the tree boughs. Some cover the substrate so thickly that there is no room for other plants. Others share habitats with liverworts, lichens, and fungi, thus providing a wonderful spectacle of hues and shapes.

Creatures associated with mosses are mostly very small. They include oval, brown mites that use their eight legs to clamber over leaves looking for any living or dead food items. They may feast on small nematodes that have thin mouth spears that are inserted into leaves to withdraw cell juices. Minute marsh-loving beetles with shiny backs covered with short bristles also occur on mosses. These beetles are so timid that not only are their heads continuously bowed, but when detected, they pull their heads into their thorax.

Also in this microcosm is the lace bug that bears a reticulated network around its body. While innocent enough in appearance, lace bugs have a beak and are not hesitant to use it, even plunging it into fingers that attempt to pick them up. Aside from the miniscule inhabitants of mosses are large male fireflies that arrive in search of females that oviposit on moss. The habitat under mosses is very suitable for the developing predaceous larvae.

Only a few examples of the more common mosses that occur in dune forests are presented here.

MOSS COMMUNITY Herbivores	 Associate
Moss mite	Firefly
Nematodes	
Minute marsh-loving beetle	
Lace bug	

DOUGLAS' NECKERA

(*Neckera douglasii*: Leucodontales: Neckeraceae)

Characteristics: Spreading moss. Branches regular, flattened, arising somewhat angular to main branch. Leaves flattened, midrib lacking.

Regional habitat: Dune forests from Washington to California.

Comments: Douglas' neckera often forms hanging mats on tree trunks and branches.

FOREST MOSS
(*Brachythecium* sp.: Hypnales: Brachytheciaceae)
Characteristics: Spreading ascending moss. Branches clustered, emerging at angle to main stem. Leaves awl-shaped, midrib present.
Regional habitat: Common in dune forests from Washington to California.
Comments: Branches of this common forest moss are held erect.

BARK MOSS
(*Polytrichum* sp.: Polytrichales: Polytrichaceae)
Characteristics: Unbranched moss. Leaves large, midrib present.

Regional habitat: Common in dune forests from Washington to California.
Comments: Bark moss grows on rocks and fallen limbs, often in combination with lichens.

ROCK MOSS
(*Racomitrium* sp.: Grimmiales: Grimmiaceae)
Characteristics: Papillose moss. Leaves narrow with slender points.
Regional habitat: Edges of dune forests from Washington to California.
Comments: Rock moss forms mats on rocks and exposed surfaces. During dry periods, the entire plant turns yellowish-brown and the leaves become crumpled, as shown here. It revives with adequate moisture.

MOSS MITE
(Oribatida: Oribatidae)
Characteristics: Adult oval, very small, dark-brown, eyeless, with thick integument. Nymph and adult stages with eight legs.
Regional habitat: Common in dune forests from Washington to California.
Comments: Moss mites are slow and clumsy. They pull in their legs, fall to the ground, and remain motionless when disturbed. They are omnivores, but prefer plant and fungal material.

MINUTE MARSH-LOVING BEETLE
(*Limnichus* sp.: Coleoptera: Limnichidae)
Characteristics: Adult small, dark brown to black, oval, shiny, antennae short, dorsum often bears erect setae; legs retractile.
Regional habitat: Damp habitats in dune forests from Washington to California.
Comments: These beetles appear to feed on the leaves of various mosses.

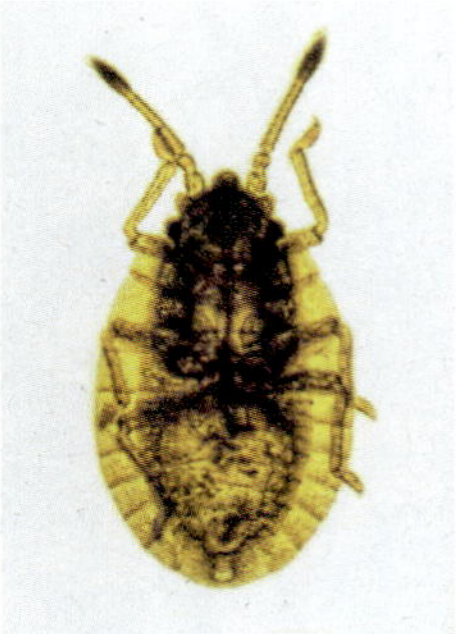

LACE BUG
(*Acalypta* sp.: Hemiptera: Tingidae)
Characteristics: Small, flattened bugs with protruding antennae. Thorax and wings covered with a shield that often has patterns that resembles lace.
Regional habitat: Common in dune forests from Washington to California.
Comments: Lace bugs use their strong beaks to imbibe juices from mosses and other plants.

FIREFLY
(*Ellychnia hatchi*: Coleoptera: Lampyridae)
Characteristics: Larva luminescent. Adult elongate, black; pronotum expanded with two red stripes; head (from above) covered by pronotum.
Regional habitat: Dune forests from Washington to California.
Comments: Both sexes are winged and non-luminescent. Larvae are carnivorous, feeding on snails and earthworms. Female fireflies frequently visit moss to deposit eggs, since larval food items gather beneath the plants.

ORCHIDS

Some of the most fascinating plants in dune forests are orchids. They add beautiful colors and bizarre shapes to the landscape. It is always a pleasant surprise to discover a blooming orchid at a turn in the path. Sometimes it is the beautiful, solitary pink flower of the fairy slipper emerging between decaying leaves. If you are close enough, you might detect its sweet fragrance wafting through the air.

If water is nearby, you have an opportunity to encounter a patch of stream orchids. The flowers are often orientated in the same direction as if they were anticipating someone's arrival or departure. The petals show an artist's blending of greens, yellows, and pinks. They make a unique color scheme with their green bracts as a backdrop.

The rattlesnake plantain orchid is somewhat hard to find, but is easily recognized by a basal rosette of leaves intricately decorated with a network of white lines. In face view, the pale yellow flowers with bulbous "cheeks" and extended "ears" resemble a monkey's face.

In wet areas in clearings or along the edges of dune forests can be found the erect stems of bog rein orchid with their vertical arrangement of white flowers. Some plants reach almost a meter in height and when several are growing together, the fragrance of their flowers is very evident.

ORCHID COMMUNITY Herbivores (including pollinators)
Bumblebees
Flower flies
Halictid bees
Mosquitoes

FAIRY SLIPPER

(*Calypso bulbosa*: Orchidaceae)

Characteristics: Small perennial orchid with one basal leaf. Flowers single; upper petals completely purple, lower lip pink with deep red lines and blotches.

Regional habitat: Shaded areas in dune forests from Washington to California.

Comments: While the distribution of this orchid, also known as calypso orchid, is almost global, it is rare in dune forests, and considered endangered in some Eastern States. It pollinates by "deception," attracting bumblebees and other insects to nectary-like structures where no nectar is produced. While searching for nectar, the orchid's pollen sacs (pollinaria) become attached to insects and are then carried to other flowers.

STREAM ORCHID

(*Epipactis gigantea*: Orchidaceae)

Characteristics: Perennial deciduous orchid. Leaves large, broad, arranged along the stem. Flowers pinkish to yellowish, arising individually from leaf axils along upper portion of stalk.

Regional habitat: Dune forests from Washington to California.

Comments: Also known as giant helleborine, this orchid is found among moss and lichens in wet habitats. Flower flies (Diptera: Syrphidae) are known pollinators.

WESTERN RATTLESNAKE PLANTAIN
(*Goodyera oblongifolia*: Orchidaceae)
Characteristics: Evergreen perennial herb. Leaves basal rosette, dark green with white network.
Regional habitat: Dune forests from Washington to California.
Comments: Flowers are said to resemble monkey faces. The common name comes from the early belief that the leaves had properties to relieve rattle-snake bite symptoms. Halictid bees are pollinators.

BOG REIN ORCHID
(*Platanthera leucostachys*: Orchidaceae)
Characteristics: Leaves closely appressed to stems. Flowers small, white; borne in many rows on erect stems.

Regional habitat: Bogs and damp areas in and around edges of dune forests from Washington to California.
Comments: Halictid bees (photo) and mosquitoes of the genus *Aedes* (Diptera: Culicidae) are pollinators.

COAST RHODODENDRON

In late spring and early summer, sections of dune forests are alive with blazes of red from the large flowers of coast rhododendron. The plants easily reach over two meters in height and often sprawl to the same width so that passing among them is difficult.

A number of insects are associated with coast rhododendron. Fragile white moths feed on the leaves, fluttering like little winged messengers under the shade of the plants while Timarcha beetles feed on the foliage. Several caterpillars also feed on the leaves. One is an extremely slender inchworm decorated with the same tints as the leaf undersurface where it feeds. This camflouge pattern is quite different from that of the stout caterpillar bearing pepper-like spots on its back.

The most curious organism living on coast rhododendron is not an insect but a fungus. This fungus sends tube-like strands (hyphae) into the leaves of the plant, thus triggering the leaves to swell and form reddish galls resembling golf balls. While a true member of the mushroom clade, this gall fungus does not form a stem or cap with gills or, for that matter, any type of fruiting body. Instead, it produces spores directly on the surface of infected gall tissue.

It is unknown whether insect herbivores or pollinators are affected by the "grayanotoxins" in coast rhododendron that cause "mad honey" disease in humans who eat rhododendron honey. Leaves of coast rhododendron are known to be poisonous to livestock.

COAST RHODODENDRON COMMUNITY	
Herbivores	**Predators-parasites**
White spotted moth	Heath gall fungus
White fly	Ladybird beetle
Slender inchworm	
Leaf beetle	
Spotted caterpillar	

COAST RHODODENDRON
(*Rhododendron macrophyllum*: Ericaceae)

Characteristics: Tall evergreen shrub. Leaves large, smooth, leathery. Flowers large, pink; borne in open clusters at tips of branches.

Regional habitat: Dune forests from Washington to California.

Comments: Also known as Pacific rhododendron and western rhododendron. The large red flowers are quite striking against a somber forest backdrop.

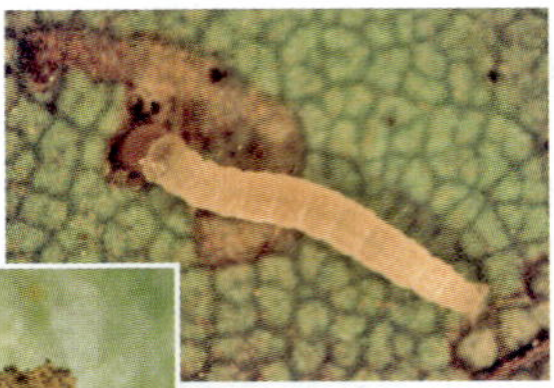

WHITE SPOTTED MOTH
(*Eucalantica polita*: Lepidoptera: Yponomeutidae)
Characteristics: Caterpillar small, body uniformly orange; head tan. Pupa light brown; forms in elliptical white cocoon usually covered by brown fecal pellets. Adult white, forewings with minute dark specks along tip, black dot in center and often at top.
Regional habitat: Dune forests from Washington to California.
Comments: Caterpillars feed fully exposed on the leaf surface. Similar moths also attack salal.

WHITE FLY
(*Aleyrodes* sp.: Hemiptera: Aleyrodidae)
Characteristics: Juvenile closely attached to leaf surface. Adult small, off-white; wings wide, front wings with faint dark blotches near tips.
Regional habitat: Dune forests in Oregon.
Comments: Adults have a weak fluttering flight. Brightly colored ladybird beetles (*Brachiacantha* sp.: Coleoptera: Coccinellidae) **(bottom)** feed on the juveniles.

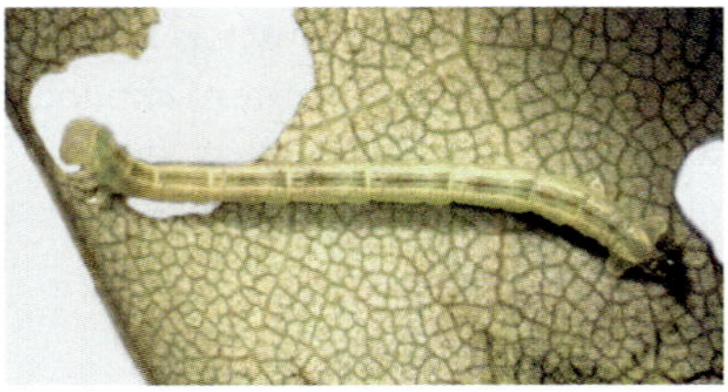

SLENDER INCHWORM
(*Dysstroma sobria*: Lepidoptera: Geometridae)
Characteristics: Caterpillar whitish with a pair of wide gray lines extending along middle of back. Adult brown; forewings with light and dark irregular bands.
Regional habitat: Dune forests from Washington to California.
Comments: Caterpillars match the color on the underside of coast rhododendron leaves.

LEAF BEETLE

(*Timarcha intricata*: Coleoptera: Chrysomelidae)

Characteristics: Larva grayish-brown, globular; pronotum and head black; legs small. Adult reddish-brown to black; head reflexed; legs relatively long; antennae bead-like; flightless.

Regional habitat: Dune forests from Washington to California.

Comments: This genus has a disjunct distribution and occurs in Western Europe as well as the Pacific Northwest. Larvae are diurnal and feed completely exposed on the leaves of coast rhododendron as well as on salal and evergreen huckleberry.

SPOTTED CATERPILLAR

(*Orthosia transparens*: Lepidoptera: Noctuidae)

Characteristics: Caterpillar body yellowish-brown with black spots and faint white mid-dorsal line; head dark brown.

Regional habitat: Dune forests in Washington and Oregon.

Comments: Caterpillars stand out against the green leaves.

HEATH GALL FUNGUS

(*Exobasidium* sp.: Exobasidiaceae)

Characteristics: Infected leaves and stems swell into roundish galls. Microscopic stalks (basidia) with four sterigmata (stalks) tipped with basidiospores develop on gall surface.

Regional habitat: Dune forests in Washington and Oregon.

Comments: This fungus, which infects coast rhododendron and other plants in the heath family (Ericaceae) does not make a fruiting body (basidiocarp) but forms spores directly on the surface of the infected tissue.

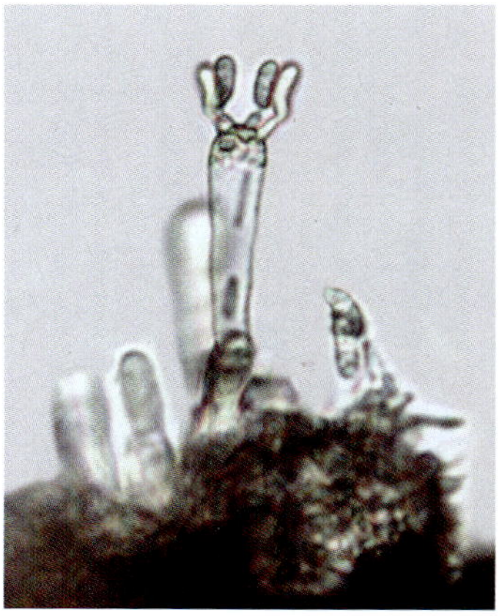

RED-FLOWERING CURRANT

The clusters of beautiful red flowers appear early in the spring, even before the leaves of red-flowering currant have completely formed. The absence of stem spines (present in closely related gooseberry plants) allows for a close-up examination of flower clusters.

Not many insects were recovered from this plant. Minute flower gnat maggots select the flower center as a secure place to develop and bore so deeply into the blooms that their white bodies are completely hidden from view. Green leaf roller caterpillars match the color of the leaves, even though they spend most of their time out of sight inside curled leaves.

RED-FLOWERING CURRANT COMMUNITY Herbivores
Flower gnat
Leaf roller caterpillar

RED-FLOWERING CURRANT
(*Ribes sanguineum*: Grossulariaceae)
Characteristics: Erect spreading perennial shrub. Leaves deciduous, alternate, five-lobed. Flowers red to pink (occasionally white); borne in clusters on flower stalks arising from leaf axils. Fruits small, round, bluish-black berries with waxy bloom.
Regional habitat: Dune forests from Washington to California.
Comments: Berries are edible but rather tasteless.

FLOWER GNAT

(*Dasineura* sp.: Diptera: Cecidomyiidae)

Characteristics: Maggot white, short, chubby with brown mouth hooks. Adult small, black, gnat-shaped.

Regional habitat: Dune forests in Washington and Oregon.

Comments: Larval development occurs inside the flowers.

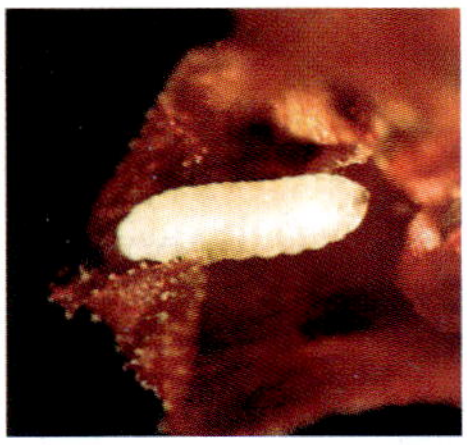

LEAF ROLLER CATERPILLAR

(Lepidoptera: Tortricidae)

Characteristics: Caterpillar yellowish-green; pronotum black; head brown.

Regional habitat: Dune forests in Washington and Oregon.

Comments: Caterpillars feed inside curled leaves.

WOOD ROSE

Except for European blackberry (found in a few disturbed areas), there are not many dune forests plants with prickles. One that does have rather strong spines is the wood rose, but it makes up for this unpleasant trait by producing large, beautiful red flowers that almost match those of coast rhododendron. The strong fragrance from the flowers is unmistakable and insects are easily attracted to it. Early Americans apparently made a tea from wood rose bark to improve their eyesight and used ground leaves to treat sore eyes.

Insect galls come in all sizes and shapes, but the most bizarre shape of all are mossy rose galls. Not even woolly bear caterpillars have hairs to match the long, stringy, red projections of these galls. The protuberances are shaped somewhat like deer antlers. Whether these outgrowths serve any purpose is unknown, but at the base of each projection is a hardened cell containing an immaculate white wasp larva. When the larva reaches a certain length, it has to start bending in order to fit within the cell. When ready to pupate, it is almost doubled over. From each pupa emerges a flawless black wasp with a chestnut-colored abdomen. However, some adults never emerge since they were attacked by an even smaller, metallic green, parasitic wasp.

Wood rose flowers are especially attractive to insects, not only to bees but also to flower flies and beetles. With so many visitors, rose flowers make an excellent abode for predaceous crab spiders that don't build webs but rest on the petals like minature crabs with their front legs spread widely, ready to grasp unsuspecting pollinators. These spiders are the epitome of patience and can remain completely motionless for hours. Once the spider has grasped its victim, there is little chance it will escape. I often wondered how wide its range of vision is with its six small eyes.

Rose hips, the mature red fruit of roses, have many uses. They are edible, made into tea, and used in cosmetics and medicine. Munching on a rose hip could give you some extra protein since there are two insects whose larvae develop inside these fruits. Both are legless and rather helpless creatures, depending on the protection provided by their food source for survival. The red rose weevil adult, with its long black beak like a thorn, blends in well when resting on a rose flower. The female uses her beak to make a hole in the rose hip for oviposition. The white larva completes it development in the hip and drops to the soil to pupate. Occasinally, a weevil larva encounters a competitor, the yellow rose hip maggot.

WOOD ROSE COMMUNITY Herbivores	Predator-parasites	Commensals
Leaf blister gall midge	Crab spider	Mites
Mossy gall wasp	Wasps	Nematodes
Flea beetle		
Petal caterpillar		
Bristly sawfly		
Smooth sawfly		
Leaf roller caterpillar		
Rose weevil		
Rose hip maggot		
Longhorn beetles		
Bruchid beetle		

WOOD ROSE
(*Rosa nutkana*: Rosaceae)
Characteristics: Small to large perennial. Leaves alternate, deciduous with five serrate oval leaflets. Flowers large, pink to red (occasionally white); usually borne singly on stalks arising from branches. Fruit (rose hips) red, roundish, containing many hairy seeds.

Regional habitat: In and adjacent to dune forests from Washington to California.

Comments: A number of insects feed on wood rose, also known as Nootka rose. Considering the strong scent of the flowers, they should be easy to locate among dune forest plants.

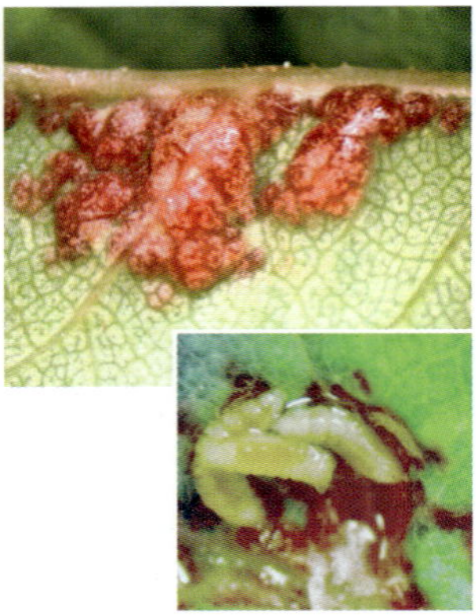

LEAF BLISTER GALL MIDGE

(*Dasineura* sp.: Diptera: Cecidomyiidae)

Characteristics: Maggots small, plump, whitish; develop inside reddish leaf blisters. Adults small, dark, with clear wings.

Regional habitat: Dune forests in Oregon.

Comments: "Blisters" extend along the mid-vein of the leaf and usually contain multiple maggots. The habitat inside the leaf blisters is quite moist.

MOSSY GALL WASP

(*Diplolepis rosae*: Hymenoptera: Cynipidae)

Characteristics: Galls resemble a red moss growing on the leaves. Larvae develop inside individual spherical cavities. Adult wasp small; head and thorax black; abdomen reddish.

Regional habitat: Dune forests from Washington to California.

Comments: Larvae of a metallic-green wasp (Hymenoptera: Torymidae) (bottom photo) parasitize the mossy gall wasp larvae.

FLEA BEETLE

(*Altica probata*: Coleoptera: Chrysomelidae)

Characteristics: Eggs whitish, elliptical, deposited on rose leaf. Larvae black, relatively elongate. Adults metallic black, shiny with straight, relatively long antennae.

Regional habitat: Dune forests from Washington to California.

Comments: Adults feed and mate on the leaf surface.

PETAL CATERPILLAR

(*Xylena* sp.: Lepidoptera: Noctuidae)

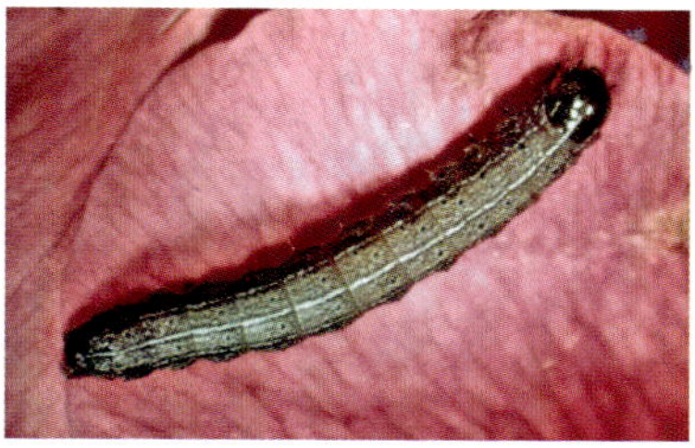

Characteristics: Caterpillar light brown; mid-dorsal line white, extending length of body; pronotum and head dark brown.

Regional habitat: Dune forests in Oregon.

Comments: Caterpillars feed almost exclusively on rose petals.

CRAB SPIDER

(*Misumena* sp.: Araneae: Thomisidae)

Characteristics: Adults mostly tan with yellowish abdomen and eight eyes; front legs held in "attack" position, somewhat like a crab.

Regional habitat: Widespread throughout dune forests from Washington to California.

Comments: Crab spiders patiently wait on flowers for hours for the arrival of pollinators. The attack consists of closing the front pair of legs around the prey, biting it and sucking out the body fluids.

BRISTLY SAWFLY

(*Cladius difformis*: Hymenoptera: Tenthredinidae)

Characteristics: Larva yellowish-green, covered with fine setae; head tan. Adult black; antennae of female straight, antennae of male with short branches; wings brown, translucent.

Regional habitat: Dune forests from Washington to California.

Comments: Larvae consume the entire leaf, not just the leaf surface.

SMOOTH SAWFLY

(*Endelomyia aethiops*: Hymenoptera: Tenthredinidae)

Characteristics: Larva yellowish-green, smooth, hairless; head tan. Adult black, elongate; legs brown; wings clear.

Regional habitat: Dune forests in Washington to California.

Comments: Larvae usually feed just on the leaf surface.

LEAF ROLLER CATERPILLAR

(*Cleptis* sp.: Lepidoptera: Tortricidae)

Characteristics: Caterpillar green; pronotum black; head brown. Adult uniformly grayish-brown; forewings with faint dark cross bands.

Regional habitat: Dune forests in Oregon and California.

Comments: Caterpillars live within folded rose leaves. They are attacked by parasitic wasps (Hymenoptera: Ichneumonidae).

ROSE WEEVIL

(*Merhynchites bicolor*: Coleoptera: Attelabidae)

Characteristics: Larva white; head capsule black. Adult deep reddish-brown; head, legs and beak black; beak straight with antennae attached near middle.

Regional habitat: Dune forests from Washington to California.

Comments: All larval development occurs within the fruit. Adults feed on leaves and petals as well as the fruit.

ROSE HIP MAGGOT

(*Rhagoletis basiola*: Diptera: Tephritidae)

Characteristics: Maggot plump, yellowish; posterior end with pair of nubbins. Adult tan; eyes reddish; wings black-striped.
Regional habitat: Dune forests from Washington to California.
Comments: Maggots feed inside the fruit, which sometimes brings them into contact with larvae of the rose weevil.

LONGHORN AND BRUCHID BEETLES

(*Pidonia* sp., *Leptura* sp.: Coleoptera: Cerambycidae) (*Bruchus* sp.: Coleoptera: Bruchidae)

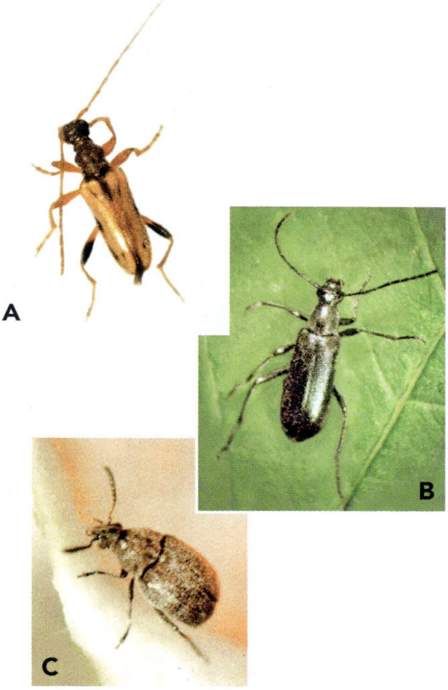

Characteristics: *Pidonia* **(A)** and *Leptura* **(B)** are medium-sized, slender longhorn beetles, with long legs and extended antennae. *Pidonia* has a dark brown head and thorax, and light brown wing covers. *Leptura* is completely metallic bluish-black. *Bruchus* **(C)** has a small, short, brown body with short legs and short antennae.
Regional habitat: Dune forests from Washington to California.
Comments: These beetles visit wood rose flowers for pollen and nectar. Longhorn beetle larvae occur inside tissues of neighboring woody plants. Larvae of *Bruchus* develop in various seeds, especially those of legumes. Mites and nematodes can be found under the elytra of the longhorn beetles.

THIMBLEBERRY

There are three common brambles in Pacific Northwest dune forests: thimbleberry, salmonberry, and trailing blackberry. They all belong to the genus *Rubus* in the rose family and are characterized by having compound fruit composed of druplets.

Large maple-shaped leaves, wide white flowers, and unarmed stems readily identify thimbleberry. The absence of spines is definitely a benefit when one is walking through thickets of thimbleberries. The large flowers with an abundance of pollen attract both bees and beetles.

THIMBLEBERRY COMMUNITY Herbivores
Stem gall wasp
Leaf roller caterpillar
Soldier beetle
Colletid bee

THIMBLEBERRY
(*Rubus parviflorus*: Rosaceae)
Characteristics: Erect unarmed perennial shrub often forming dense thickets. Leaves large, maple-shaped with finely hairy surfaces; deciduous. Flowers distinct, white; borne singly or in small groups at tips of short stalks emerging from branches. Fruit red, flask-shaped.
Regional habitat: Dune forests from Washington to California.
Comments: Fruits are edible but not especially tasty.

STEM GALL WASP

(*Diastrophus kincaidii*: Hymenoptera: Cynipidae)

Characteristics: Larva small, white; legless; develops inside gall. Adult small, black; legs yellow; antennae long, constantly moving; wings fuscous.

Regional habitat: Dune forests from Washington to California.

Comments: The swollen galls containing larvae are located on the stems of thimbleberry.

LEAF ROLLER CATERPILLAR

(*Acleris keiferi*: Lepidoptera: Tortricidae)

Characteristics: Caterpillar greenish with black head. Adult with white- and brown-banded forewings.

Regional habitat: Dune forests from Washington to California.

Comments: Caterpillars live inside folded leaves.

SOLDIER BEETLE

(*Podabrus cavicollis*: Coleoptera: Cantharidae)

Characteristics: Adult brownish, medium-sized; thorax roundish; tip of abdomen protruding beyond wing covers; legs brown.

Regional habitat: Dune forests from Washington to California.

Comments: These beetles will often feed for long periods on pollen in the center of thimbleberry flowers.

COLLETID BEE

(*Hylaeus* sp: Hymenoptera: Colletidae)

Characteristics: Adult small, black, with large head; antennae short and erect; wings fuscous.

Regional habitat: Dune forests from Washington to California.

Comments: Colletid bees often share thimbleberry flowers with soldier beetles.

SALMONBERRY

From Alaska to California, the red flowers of salmonberry are like beacons announcing the arrival of spring. Salmonberry stems, which can reach 4 meters in length, have only a few thorns so clothing is not constantly snagged as you pass through their thickets. Northwest coast peoples ate salmonberry sprouts and berries. Its fruit was often served with salmon, which is how the plant received its name. While ripe berries can be quite tasty, they are soft, and when I reached home after picking some, they had become a crushed, wet compote in my pocket.

Many insects, from aphids, flies, and beetles to moths and sawflies, find salmonberry a choice food plant. Two species of leaf-feeding sawfly larvae reveal different survival tactics. One is covered with rows of forked spines which must provide some protection from predators. The other releases a black droplet, presumably some type of repellent, from the back of its head when disturbed.

The numerous caterpillars feeding on salmonberry leaves often are attacked by internal fly maggots. One way to avoid this fate is to become a leaf miner, and develop only within the leaf between the epidermal layers. To fit in such a confined space, the caterpillars must be small and flat. It sounds like a perfect, safe environment. However, little wasps search out and deposit their eggs into the caterpillars through the leaf epidermis.

A little fly forms mines quite similar to those of the caterpillar; however, the legless maggots are gregarious with usually three or four found in each mine. These maggots scrape the plant matter into their guts with mouth hooks, as if they were hoeing out a patch of weeds.

A number of other invertebrates visit salmonberry, one of which is the banana slug. Banana slugs are the second largest slug in the world, easily reaching 225 mm in length. Feeding high on salmonberry stems protects them from raccoons that avidly seek out these molluscs.

An assortment of beetles occurs on salmonberry leaves, including large click-beetles that drop to the ground when disturbed. If they happen to land on their backs, they can force the front part of their body against the hind part, which propels them into the air with a 50–50 chance they will land right-side up. If not, they simply keep clicking until they land on their feet.

The salmonberry leaf beetle is on the brink of extinction and probably will soon disappear in most dune forests. It is one of the most passive beetles I have encountered, feeding completely exposed on leaves day and night. And enemies abound. A little fly maggot punctures the larva's skin and slowly drains its blood. There seems to be nothing the leaf beetle can do to stop or prevent the assault, which is usually fatal. A small bug with a long beak also samples the beetle's blood. Even if the beetle survives the attacks from these two assailants, it could die from viral infections. The attacks don't end when the beetle larvae enter the soil to pupate since parasitic nematodes lurk underground.

These microscopic worms enter and kill both larvae and pupae. Eventually hundreds of infective stage juvenile nematodes swarm out of the beetle cadavers.

Such formidable predation may be why these leaf beetles are so rare. When the obstacles against survival are overwhelming, extinction is inevitable.

SALMONBERRY COMMUNITY Herbivores	Predators	Parasites
Spiny sawfly	Soldier beetle	Heterorhabditid nematodes
Black-headed sawfly	Red-backed spider	Gregarine protozoans
Obscure root weevil	Velvet mite	Poxvirus
White-ribbon moth	Flower fly	Tachinid fly
Tan mottled moth		Braconid wasps
Scribe moth		Gall gnat larva
Carpet moth		Red-eyed bug nymph
Red-striped looper caterpillar		
Mottled leaf roller caterpillar		
Red leaf roller		
Leaf miner		
Leaf beetle		
Click beetle		
Berry aphid		
Leaf mining fly		
Flat millipede		
Ant-mimic bug		

SALMONBERRY
(*Rubus spectabilis*: Rosaceae)
Characteristics: Erect tall spreading perennial, often forming dense thickets. Leaves with three serrate leaflets. Flowers large, borne singly or in pairs at the tips of short stem branches, with five pink to red petals and clusters of central stamens; fruit a yellowish to reddish berry.
Regional habitat: Dune forests from Washington to California.
Comments: The berries are eaten by a variety of birds and mammals.

SPINY SAWFLY
(*Monophadnoides rubi*: Hymenoptera: Tenthredinidae)
Characteristics: Larva light green, covered with forked spines; head black. Adult shiny, black; legs yellow; wings dark, translucent.
Regional habitat: Dune forests from Washington to California.
Comments: The forked spines covering the body of the larvae are unique and readily identify this species.

BLACK-HEADED SAWFLY

(*Priophorus brullei*: Hymenoptera: Tenthredinidae)

Characteristics: Larva with light green sides and a wide, dark, dorsal stripe; head brown.

Regional habitat: Dune forests from Washington to California.

Comments: When disturbed, the larva emits a drop of black liquid from the back of its head or neck.

OBSCURE ROOT WEEVIL

(*Sciopithes obscurus*: Coleoptera: Curculionidae)

Characteristics: Adult grayish-brown; abdomen wide; beak short; antennae long.

Regional habitat: Dune forests from Washington to California.

Comments: Eggs are laid in a folded edge of a leaf. Larvae feed on the roots of salmonberry and other plants. Adults eat the foliage and flowers.

WHITE-RIBBON MOTH

(*Mesoleuca gratulata*: Lepidoptera: Geometridae)

Characteristics: Caterpillar light green with faint white lines and rows of dark green protuberances. Adult with alternating white and dark brown irregular bands on the forewings.

Regional habitat: Dune forests from Washington to California.

Comments: Caterpillars feed on leaves of salmonberry.

TAN MOTTLED MOTH

(*Herpetogramma pertextalis:* Lepidoptera: Crambidae)

Characteristics: Caterpillar various shades of green; each segment with a circle of eight rings; head tan. Adult light brown with mottled wings.

Regional habitat: Dune forests from Washington to California.

Comments: Coloration and position of the rings on the caterpillars varies with age.

SCRIBE MOTH

(*Habrosyne scripta*: Lepidoptera: Drepanidae)

Characteristics: Caterpillar large, bulky, gray to brown, with black mid-dorsal line; can have one or two white blotches behind head or additional white spots on other body parts. Pupa reddish-brown, often formed within portion of last larval skin. Adult brown to gray; forewings with intricate white, often zigzag lines.

Regional habitat: Dune forests from Washington to California.

Comments: There is considerable variation in the color patterns of both adults and caterpillars. Fly maggots (Diptera: Tachinidae) (bottom photo) parasitize the caterpillars, eventually emerging as white maggots that quickly form reddish puparia. Adult flies usually emerge within two weeks after pupation.

CARPET MOTH

(*Dysstroma* sp.: Lepidoptera: Geometridae)

Characteristics: Caterpillar long, slender; yellowish-green with widely spaced white rings; head tan. Pupa yellowish-greenish; abdomen often curved. Adult dark brown; forewings with reddish-brown irregular bands.

Regional habitat: Dune forests from Washington to California.

Comments: The caterpillars are parasitized by maggots (Diptera: Tachinidae) that form robust brown puparia after emerging **(A)**. Red-striped looper caterpillars (Geometridae) **(B)** often occur together with carpet moth caterpillars.

MOTTLED LEAF ROLLER CATERPILLAR

(*Epinotia* sp.: Lepidoptera: Tortricidae)

Characteristics: Caterpillar light green, with dark mid-dorsal line running length of body; head grayish with central black spot. Pupa with anterior part green, posterior portion brown. Adult light grayish-brown; forewings with darker brown patterns.

Regional habitat: Dune forests from Washington to California.

Comments: Another caterpillar on salmonberry is the red leaf roller (*Acleris* sp.: Tortricidae) **(A, B)**. The brownish caterpillars consume berries. The adult moths are uniformly reddish-brown.

LEAF MINER

(*Coptotriche aenea*: Lepidoptera: Tischeriidae)

Characteristics: Caterpillar small, flat; greenish-yellow; thorax with swollen area; legs reduced. Adult small; forewings uniformly dark brown.

Regional habitat: Dune forests in Oregon.

Comments: The caterpillars, which are structurally modified for movement between the epidermal leaf layers, form blotch or linear mines on the leaves of salmonberry. Small wasps (Hymenoptera: Braconidae) deposit eggs in the leaf mines and, after hatching, the wasp larvae parasitize the leaf miner caterpillars.

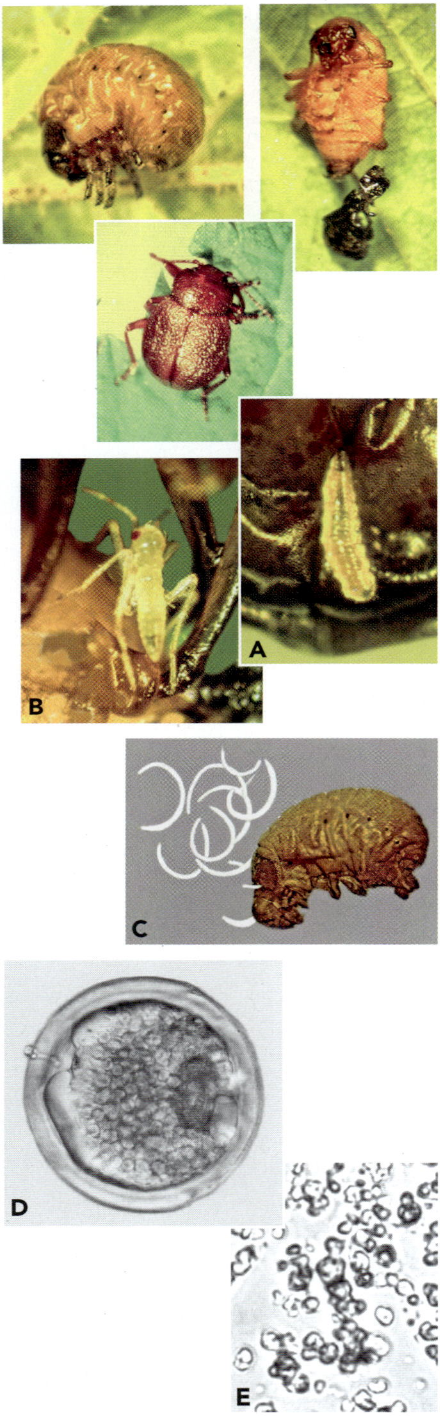

LEAF BEETLE

(*Timarcha cerdo*: Coleoptera: Chrysomelidae)

Characteristics: Larva reddish-brown; body C-shaped; head bent down. Pupa light brown; formed in soil. Adult reddish-brown, stocky; head bent downward; antennae straight, bead-like; elytra fused.

Regional habitat: Dune forests from Washington to California.

Comments: This is a rare beetle on the verge of extinction. The larvae feed completely exposed during the day. While their blood is considered to be toxic, the larvae are still attacked by predators, parasites, and pathogens, all of which contribute to their decline. Ectoparasites include small gall gnat larvae (Diptera: Cecidomyiidae) **(A)** and red-eyed bug nymphs (Hemiptera) **(B)** that imbibe blood through the larvae's cuticle. Parasitic nematodes (*Heterorhabditis* sp.: Nematoda) **(C)** penetrate and multiply in the body cavities of mature larvae that enter the soil for pupation. Protozoans (Apicomplexa: Eugregarinorida) **(D)** and poxviruses (Poxviridae) **(E)** also infect the larvae.

CLICK BEETLE

(*Agriotes* sp.: Coleoptera: Elateridae) **(A)**

SOLDIER BEETLE

(*Silis* sp.: Coleoptera: Cantharidae) **(B)**

Characteristics: Click beetle: thorax large; pronotum with extended hind angles; elytra mostly black but reddish at base. Soldier beetle: pronotum red; head black; antennae as long as body; elytra brownish.

Regional habitat: Dune forests from Washington to California.

Comments: The click beetle feeds on leaves of salmonberry while the soldier beetle searches for pollen and small insects.

BERRY APHID

(*Amphorophora* sp.: Hemiptera: Aphididae)

Characteristics: Adult slender, yellowish-green; eyes small, red; antennae extended; cornicles elongate.

Regional habitat: Dune forests from Washington to California.

Comments: Feeds on salmonberry and other species of *Rubus*. Aphid populations can build up in the summer months. They are a favorite food for flower fly larvae (Diptera: Syrphidae), the adults of which can be found resting on salmonberry leaves (right photo).

LEAF MINING FLY

(*Pegomyia* sp.: Diptera: Anthomyiidae)
Characteristics: Maggot fleshy, white to yellowish; mouth hooks black. Puparium reddish-brown, cylindrical, posterior end with two short projections. Adult small, gray; eyes large; antennae short; wings clear.
Regional habitat: Dune forests from Washington to California.
Comments: The maggots form communal blotch mines on salmonberry leaves. Pupation occurs within the mines.

ANT-MIMIC BUG

(*Orectoderus* sp.: Hemiptera: Miridae)
Characteristics: Adult slender; abdomen constricted at connection with thorax; antennae long, straight, backward positioned.
Regional habitat: Dune forests from Washington to California.
Comments: Ant-mimic bugs are herbivores and obviously obtain some protection from mimicking biting or stinging worker ants.

FLAT MILLIPEDE
(*Diplopoda*: Polydesmidae)
RED-BACKED SPIDER
(*Enoplognatha ovata*: Theridiidae)
VELVET MITE
(*Acari*: Thrombidiidae).

These represent non-insect invertebrates that occur on salmonberry leaves.

Characteristics: Flat millipede **(A)**: brown, narrow, flattened; most segments bearing two pairs of legs. Red-backed spider **(B)**: abdomen usually reddish (pattern variable); legs long. Adult velvet mite **(C)**: red, chubby; eight orange legs.

Regional habitat: Dune forests from Washington to California.

Comments: The flat millipede is an herbivore. The red-backed spider, that constructs flimsy webs on leaves, and the velvet mite are predators.

TRAILING BLACKBERRY

Instead of forming an erect plant, trailing blackberry sends thin, horizontal, wiry, armed stems in all directions, between and over neighboring plants. Walking near these plants is hazardous, since the ropey vines, that root at the nodes, cross forest paths and are easy to trip over. Large leaves make it difficult to detect the presence of the prostrate stems.

TRAILING BLACKBERRY COMMUNITY Herbivores
Inchworm caterpillar
Green sawfly

TRAILING BLACKBERRY
(*Rubus ursinus*: Rosaceae)
Characteristics: Prostrate, prickly, trailing deciduous perennial, often growing over neighboring vegetation. Leaves with three, large, toothed leaflets. Flowers mostly white (sometimes pinkish); borne in small clusters at ends of stalks rising from leaf axils. Male and female flowers separate. Fruits black, edible.

Regional habitat: Dune forests from Washington to California.
Comments: Trailing blackberry is not to be confused with the introduced Asian Himalayan blackberry (*Rubus discolor*), a bushy plant that occasionally occurs in disturbed areas along the edges of dune forests.

INCHWORM CATERPILLAR

(Lepidoptera: Geometridae)

Characteristics: Caterpillar long, slender; yellow-brown with dark brown diamond pattern running along back; head black. Pupa brown with pointed tip. Adult pale gray; wings with darker gray bands.

Regional habitat: Dune forests in Oregon.

Comments: Both caterpillar, feeding on the leaves, and adult appear to be undescribed.

GREEN SAWFLY

(*Monophadnoides* sp.: Hymenoptera: Tenthredinidae)

Characteristics: Larva light green; head tan; eyes dark.

Regional habitat: Dune forests in Oregon.

Comments: The green larvae are well camouflaged as they feed on the leaves.

STARRY FALSE SOLOMON'S SEAL

Long broad alternating leaves on a single stem with small white flowers borne at the tip are identifying features of starry false Solomon's seal. Similar size and shape of the alternate leaves resembles a ladder. Fewer than 12 flowers in the terminal raceme and bluish-black mature berries separate this species from large false Solomon's seal (*Smilacina racemosa*) that has 20 or more flowers and reddish mature berries. The black-spotted sawfly larva that feeds on the leaves closely resembles those found on the leaves of snakeberry (*Maianthemum dilatatum*).

STARRY FALSE SOLOMON'S SEAL COMMUNITY **Herbivore**
Black-spotted sawfly larva

STARRY FALSE SOLOMON'S SEAL

(*Maianthemum stellatum*: Asparagaceae)

Characteristics: Erect perennial herb with stem emerging from buried rhizomes. Leaves elliptical, alternate, bases clasping, veins parallel; arranged in two rows along stem. Flowers white, star-like; 12 or fewer borne in terminal cluster. Fruit light green then becoming blue-black berry.

Regional habitat: Dune forests from Washington to California.

Comments: This little plant has a wide distribution throughout North America.

BLACK-SPOTTED SAWFLY LARVA

(*Phymatocera* sp.: Hymenoptera: Tenthredinidae)

Characteristics: Larva faint-yellowish with black pepper spots; head black.

Regional habitat: Dune forests from Washington to California.

Comments: Based on appearance, this may be the same or a closely related species that occurs on snakeberry (*Maianthemum dilatatum*).

STEEPLEBUSH

The long pink flower spikes of steeplebush are found around the edges of dune forests. The individual flowers are clustered together so tightly that it is impossible to have a clear view of any individual bloom. With six-foot stems, steeplebush makes an impressive sight, especially when the plants are growing in dense, virtually impenetrable groups.

Some branch tips had their leaves clustered together. After carefully removing all the leaves from one branch, I discovered a small gall made of compressed leaves and stem tissue. Inside the gall was a small menagerie, a virtual collection of miniscule insects living together in apparent harmony. One little gall gnat starts it off by laying an egg in the plant stem. Steeplebush responds by forming a series of misshapen leaves around the point of attack. This provides food for the little gall gnat maggot and for another type of gall gnat maggot that suddenly arrives on the scene. Little predatory wasp larvae next appear, devouring many of the gall gnat maggots. Maggots that do survive pupate and emerge as small delicate gnat-like flies. The Northwest coast peoples used branched twigs of steeplebush to rake tooth shells from the sea bottom. These shells were used for adornment and trade.

It is interesting how various caterpillars feeding on the same plant use different survival tactics. A tan-colored inchworm mimics twigs. Colored like a twig, the caterpillar will remain motionless for long periods with its posterior legs grasping the twig, its body bent over, and its tiny forelegs withdrawn. A brown-spotted caterpillar doesn't seem very well camouflaged but still survives. The green shades caterpillar matches the leaf color, but then has two white spots that should make it very visible. The slender caterpillar with a grayish-brown back bearing a number of orange cross bars is another colorful anomaly. Perhaps we have to look through the eyes of a predator to comprehend which prey is visible and which is not.

It is interesting how colors on insects are intermediate between the hues we use for descriptions. For instance, the color on the back of the two-banded caterpillar varies between brown and gray, that on the brown-spotted caterpillar between yellow and tan, and the color on the back of the three-lined caterpillar is between brown and purple.

STEEPLEBUSH COMMUNITY Herbivores	Parasite	Inquiline
Tan inchworm	Wasps	Gall gnat
Brown-spotted caterpillar		
Green angle shades caterpillar		
Two-banded caterpillar		
Three-lined caterpillar		
Leaf gall gnats		

STEEPLEBUSH

(*Spiraea douglasii*: Rosaceae)

Characteristics: Erect shrub. Leaves alternate, green above, pale beneath, toothed along upper half. Flowers small, pink to rose; crowded together in tight cone-shaped clusters at stem tips. Fruit tiny, dry.

Regional habitat: Dune forests from Washington to California.

Comments: Caterpillars found on the leaves display a wide variation in color and markings, making identification difficult.

TAN INCHWORM

(*Eulithis* sp.: Lepidoptera: Geometridae)

Characteristics: Caterpillar long, slender, tan with dark markings on back; posterior legs well developed for clasping stems, anterior legs minute.

Regional habitat: Dune forests in Oregon.

Comments: When disturbed during feeding, the caterpillar rises up on its hind legs, forms an arch and becomes motionless, thus mimicking a twig of steeplebush.

BROWN-SPOTTED CATERPILLAR

(*Orthosia* sp.: Lepidoptera: Noctuidae)

Characteristics: Caterpillar large, fleshy, yellow with brown spots; head and pronotum reddish-tan.

Regional habitat: Dune forests in Oregon.

Comments: Color and markings vary depending on age.

GREEN ANGLE SHADES CATERPILLAR

(*Phlogophora periculosa*: Lepidoptera: Noctuidae)

Characteristics: Caterpillar fleshy; greenish with faint, dark midline running length of body; subterminal segment with pair of white spots.

Regional habitat: Dune forests in Oregon.

Comments: The white spots resemble eyes. Color and markings vary depending on age.

TWO-BANDED CATERPILLAR

(*Lacanobia* sp.: Lepidoptera: Noctuidae)

Characteristics: Caterpillar slender, grayish-brown, with series of transverse orange bars on back; sides with dark brown bands running length of body; head tan.

Regional habitat: Dune forests in Oregon.

Comments: Color and markings vary depending on age.

THREE-LINED CATERPILLAR

(*Xylena* sp.: Lepidoptera: Noctuidae)

Characteristics: Caterpillar gray-brown, covered with black spots; three white lines running length of body; head black.

Regional habitat: Dune forests in Oregon.

Comments: Color and markings vary depending on age.

LEAF GALL GNATS

(*Parallelodiplosis* sp. and *Contarinia* sp.: Diptera: Cecidomyiidae)

Characteristics: Maggots small, white; legless; living inside leaf gall tissue. Adults small, dark; legs long; wings transparent.

Regional habitat: Dune forests from Washington to California.

Comments: Gall inhabitants can be many and varied. One gall gnat (probably *Parallelodiplosis* sp.) initiates the gall and deposits eggs and then a second gall gnat (*Contarinia* sp.) (bottom image) arrives to deposit its eggs after the gall has formed. The second species is considered an inquiline. Parasitic wasps deposit eggs in the gall and their larvae prey on the gall gnat maggots.

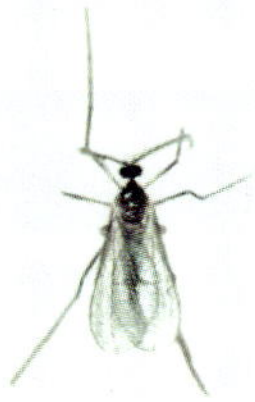

WESTERN HEMLOCK

With trunks stretching upwards 70 meters, lofty branches bearing slender twigs with flat needles, and short, persistent cones, western hemlock is a most impressive dune forest tree. It is not surprising that western hemlock, which can reach an age of 1000 years, is the official State Tree of Washington. This conifer is well adapted to Pacific Northwest forests and is more shade tolerant than Douglas-fir and Sitka spruce.

The associates of western hemlock are underestimated here simply because it was difficult to obtain foliage for examination with most of the branches being out of reach. Of the insects found, the little needle binder caterpillar was most interesting since it protects itself by constructing a shelter out of hemlock needles. Without its case, the caterpillar is quite noticeable, but only some slight movements of the case indicate its presence within.

Some white, fuzzy material hanging down from the twigs of western hemlock turned out to be small black aphids, better known as hemlock woolly adelgids. The fuzzy material or "wool" is actually a series of egg sacks with newly hatched young. This tiny insect, which was introduced from Asia, has become a serious pest of hemlocks throughout the United States. Since applications of pesticides can contaminate adjacent aquatic habitats, attempts have been made to use biological agents, such as predatory beetles and fungi, for control. Perhaps the fungus I found developing inside a dead adelgid larva is one of those being used to control these pests.

Conks are large, hard, fruiting bodies of fungi that are usually attached to woody parts of trees. The mycelial portion of these conks lives inside trees, often causing disease. A conk found on a dead log of western hemlock could have killed the tree. It was as large as a dinner plate and looked soft and pliable, but was actually quite hard. A little brown caterpillar selects conks as a food supply, eating its way through the woody pulp. While the caterpillar is able to eat through the conk, the moth is not. That is why, before it pupates, the caterpillar forms a little chimney out of conk particles so the pupa can wiggle through it before the brown-spotted moth emerges. The caterpillar shares this coarse diet with a little midge larva and a burrowing beetle.

WESTERN HEMLOCK COMMUNITY		
Herbivores	Detritivore	Parasite-pathogen
Woolly adelgid	Cyanide millipede	Sulfur shelf polypore
Needle miner		
Needle binder		
Sawfly		
Polypore midge		
Polypore moth		
Polypore burrowing beetle		

WESTERN HEMLOCK

(*Tsuga heterophylla*: Pinaceae)

Characteristics: Tall tree with a rough bark and narrow crown. Needles single, short, flat, with blunt tips, green on top, two white stripes below; attached to twigs by stubby bases that persist after needles fall. Cones small, brown, without prickles.

Regional habitat: Dune forests from Washington to California.

Comments: The genus occurs in Asia and based on fossils, at one time thrived in Europe.

WOOLLY ADELGID

(*Adelges tsugae*: Hemiptera: Adelgidae)

Characteristics: Adult female stationary, deposits white cottony egg tufts that adhere to needles and twigs. Larvae small, dark, very mobile, usually covered with cottony wax particles.

Regional habitat: On hemlocks in dune forests from Washington to California.

Comments: This insect was introduced from Asia. Some individuals were infected by a fungus with septate mycelium and oval spores (bottom right).

NEEDLE MINER

(*Coleotechnites* sp.: Lepidoptera: Gelechiidae)

Characteristics: Caterpillar small, greenish-gray with brown head. Pupa slender, dark brown. Adult small, grayish, with narrow mottled wings.

Regional habitat: Dune forests from Washington to California.

Comments: The caterpillars develop on both sides of the needle's midrib, leaving behind their fecal matter. Pupation occurs inside the hollowed out needles or in debris beneath the tree.

SAWFLY

(*Neodiprion tsugae*: Hymenoptera: Diprionidae)

Characteristics: Larva green. Cocoon reddish-brown. Adult brownish-black with tan wings. Male antennae feathery (shown).

Regional habitat: Dune forests from Washington to California.

Comments: Larvae feed on needles.

NEEDLE BINDER

(*Argyrotaenia* sp.: Lepidoptera: Tortricidae)

Characteristics: Caterpillar reddish-brown with a tan head and a dark-brown dorsal stripe running length of body.

Regional habitat: Dune forests from Washington to California.

Comments: The behavior of the needle binder caterpillar varies. Some caterpillars feed within a shelter made by webbing together adjacent growing green needles. Other caterpillars use dead needles to form an enclosed shelter.

SULFUR SHELF POLYPORE

(*Laetiporus conifericola*: Polyporales: Fomitopsidaceae)

Characteristics: Fruiting bodies (basidiocarps) pink to reddish (shown), spores released from many small pores on lower surface.

Regional habitat: Dune forests from Washington to California.

Comments: This fungus attacks a number of conifers, but western hemlock in dune forests is especially susceptible.

POLYPORE MOTH

(*Decantha* sp.: Lepidoptera: Oecophoridae)

Characteristics: Caterpillar grayish with tan to brown head. Pupa brown. Adult small, with mottled brown and white forewings with rounded tips.

Regional habitat: Dune forests from Washington to California.

Comments: The caterpillars, which are difficult to find, bore into the polypore and make tunnels as they feed. The pupae wiggle up to protrude from the larval feeding gallery making it easier for adult moths to emerge. This is one of several small moths that feed in conks and polypores.

POLYPORE MIDGE

(*Chironomus* sp.: Diptera: Chironomidae)

Characteristics: Larva slender, elongate, with irregular transverse bars along length of body. Adult small, dark, with long legs and transparent wings. Males have feathery antennae.

Regional habitat: Dune forests in Oregon.

Comments: Larvae feed on fungal tissue on the underside of the polypore.

POLYPORE BURROWING BEETLE

(*Endecatomus* sp.: Coleoptera: Endecatomidae)

Characteristics: Larva whitish, develops in polypore. Adult minute, brown, with reflexed head and short, thin legs.

Regional habitat: Dune forests from Washington to California.

Comments: These very small beetles also occur in other woody fungi.

CYANIDE MILLIPEDE

(*Harpaphe haydeniana*: Xystodesmidae)

Characteristics: Bluish-black many-segmented body with dorsal segments bearing yellow tips; antennae black.

Regional habitat: Dune forests along the Pacific Northwest from Washington to California.

Comments: These blind millipedes, which feed on leaf litter, can emit hydrogen cyanide as a defense reaction. The distinct contrasting black-yellow color pattern is an example of aposematic coloration that warns predators to leave them alone.

HUCKLEBERRY

Thick green patches of evergreen huckleberry are commonly encountered in Pacific Northwest dune forests. The little, bell-shaped, pinkish-white flowers look like those of blueberry and, if you are lucky, you might catch a glimpse of the beautiful indigo butterfly flittering around the flower clusters. It just stops to sample nectar since its eggs are laid on other plants. Not many butterflies visit dune forests since conifers have little to offer as rewards.

Two quite different herbivorous insects often occur together on the leaves of evergreen huckleberry. The little leaf blotch miner moth caterpillar busily digests the center of the huckleberry leaves, using the upper and lower leaf epidermis as roof and floor of its abode. It eventually pupates inside its residence. A little gall gnat maggot searches out the same dwelling place since it provides food and protection. Both insects share the same mine and appear to tolerate each other.

Here again we meet the little Timarcha leaf beetle. The European Timarcha beetle releases a toxic secretion from its leg joints and mouth when irritated. This reflex bleeding is not found in the American species when irritated. Instead, they lower their heads and release droplets of excretia from their digestive tract. Whether this excrement has some toxic properties that repel predators is not known.

One day, as I rounded a curve in the path, I noticed a very strange deposit on a number of small stems of evergreen huckleberry. It looked like someone had attached red clay to the stems, giving the impression of a group of candles with the tips of the huckleberry stems as wicks. It turned out to be a parasitic fungus! I had never seen huckleberry rust before and did not realize that these fungal deposits slowly strangle the plants, cutting off their supply of moisture and minerals. Some of the stems protruding above the fungal sheaths had already died. But only part of the rust's life cycle is on evergreen huckleberry. The other part of the cycle occurs on conifers, the fungus alternating back and forth between these two quite different hosts.

Locating the much less aggressive red huckleberry, which prefers shade with intermittant periods of sun, is difficult. Sometimes, when looking for this plant, it is easier to search for stems with red leaf buds, which are actually galls inhabited by a little gall gnat. A brighter red bud indicates that a second type of maggot has made its home there, too. Also on red huckleberry is a caterpillar that binds the leaves together with silk and rarely leaves its shelter. The maggots of still another fly mine the leaves.

These insects on both species of huckleberry, as well as many others on other dune forest plants, show how a single plant can be attacked by multiple insects with each insect finding its own niche that rarely overlaps with that of other insects. Even when the same niche is occupied, temporal separation of the inhabiting insects often occurs, with some appearing in early spring and others in mid-summer or late fall. How many fierce competitions occurred before these balances were established is impossible to say.

EVERGREEN HUCKLEBERRY COMMUNITY Herbivores	Parasite
Indigo butterfly	Huckleberry rust
White-striped caterpillar	
Leaf blotch miner moth	
Gall gnat maggot	
Leaf tip roller	
Leaf beetle	

EVERGREEN HUCKLEBERRY
(*Vaccinium ovatum*: Ericaceae)
Characteristics: Erect evergreen bush. Leaves small, shiny, finely serrate, alternate, leathery. Flowers pinkish-white, bell-shaped; borne in clusters arising from leaf axils. Mature fruit a deep purplish-black berry.

Regional habitat: Dune forests from Washington to California.
Comments: Evergreen huckleberry can grow up to two meters in height in well-shaded forests but is usually not over a meter high in partially exposed areas. The berries are flavorful when ripe.

INDIGO BUTTERFLY

(*Lycaeides* sp.: Lepidoptera: Lycaenidae)

Characteristics: Wings silvery, violet-blue; posterior margin of hindwings with row of white circles with black spots. Female forewings with diffuse black borders.

Regional habitat: Dune forests from Washington to California.

Comments: The host plants of the caterpillars are unknown since only the adults were seen visiting the flowers of evergreen huckleberry.

WHITE-STRIPED CATERPILLAR

(*Aseptis* sp.: Lepidoptera: Noctuidae)

Characteristics: Caterpillar green with broad lateral white stripes and dorsal and sub-dorsal white lines running length of body; head tan with dark spots.

Regional habitat: Dune forests from Washington to California.

Comments: The caterpillars eat portions of the leaves. The white lateral stripes would appear to attract predators and parasites.

LEAF BLOTCH MINER MOTH

(*Cameraria nemoris*: Lepidoptera: Gracillariidae)

Characteristics: Early stage larva yellowish, flat; head triangular, protruding. Later instars more caterpillar-like. Pupa yellowish-brown **(A)**; formed under sheet of webbing inside leaf mine. Adult very small, light brown; forewings with white bands.

Regional habitat: Dune forests from Washington to California.

Comments: The caterpillars form blotch mines on the leaves. The mines often contain gall gnat maggots (Diptera: Cecidomyiidae) **(B)** that live alongside the leaf blotch miner caterpillars as inquilines. Closely related leaf blotch miners occur on salal and twinberry. Associations with gall gnats also occurs with leaf blotch miners on twinberry.

LEAF TIP ROLLER

(*Olethreutes* sp: Lepidoptera: Tortricidae)

Characteristics: Caterpillar yellowish with a black head. Adult mottled brown and white with two faint, wide, darker bands across forewings.

Regional habitat: Dune forests from Washington to California.

Comments: The caterpillars bind together terminal leaves and use them as shelters.

LEAF BEETLE

(*Timarcha intricata*: Coleoptera: Chrysomelidae)

Characteristics: Larva short, broad, light brown with black head. Adult shiny reddish-brown to black with reflexed head, long legs and long, beaded antennae; elytra fused.

Regional habitat: Dune forests from Washington to California.

Comments: Larvae feed completely exposed during the day. While their blood is considered to be toxic, they are still attacked by predators and parasites. This leaf beetle also occurs on salal and coast rhododendron.

HUCKLEBERRY RUST

(*Pucciniastrum goeppertianum*: Pucciniales: Pucciniastraceae)

Characteristics: Huckleberry rust forms thick reddish rinds around stems of evergreen huckleberry. When fungal spores mature and are ready for release, the rind becomes dry and brittle.

Regional habitat: Dune forests from Washington to California.

Comments: By withdrawing nourishment, the fungus can kill the stem or damage it so that the protruding portion only produces small, frail leaves. This rust has two hosts (demicyclic). On evergreen huckleberry it produces one type of spore (teliospores) and on firs it produces a second spore type (aeciospores). This alternation of plant hosts for completion of the fungus life cycle is typical for rust fungi.

RED HUCKLEBERRY COMMUNITY Herbivores	Parasite
Bud gall gnat	Wasps
Leaf binder caterpillar	
Blotch miner fly	

RED HUCKLEBERRY
(*Vaccinium parvifolium*: Ericaceae)
Characteristics: Erect deciduous bush with green, strongly angled branches. Leaves small, oval, thin, alternate, entire. Flowers tiny, pinkish-red, bell-shaped; borne singly on stalks arising from leaf axils. Mature fruit a deep red berry.

Regional habitat: Dune forests from Washington to California.
Comments: Red huckleberry leaves are quite delicate in contrast to the thick, leathery leaves on evergreen huckleberry.

BUD GALL GNAT

(*Dasineura* sp.: Diptera: Cecidomyiidae)

Characteristics: Maggot small, white, broad; lives inside enlarged leaf bud that turns deep red. Adult small, dark with long antennae; wings two-veined.

Regional habitat: Dune forests in Oregon.

Comments: The striking color change of the bud after infestation makes this association easy to detect.

LEAF BINDER CATERPILLAR

(*Pyla* sp.: Lepidoptera: Pyralidae)

Characteristics: Caterpillar yellowish with dark head; feeds inside silk-bound leaves. Pupa slender, tan; forms within caterpillar leaf shelters. Adult forewings various shades of mottled white and tan with brown transverse bands.

Regional habitat: Dune forests in Oregon and California.

Comments: The caterpillar shown here was lethargic, swollen, and contained a number of developing parasitic wasp larvae (Hymenoptera: Encyrtidae).

BLOTCH MINER FLY

(*Phytomyza* sp.: Diptera: Agromyzidae)

Characteristics: Maggot white with pointed head. Adult black, resembling small housefly; wings clear.

Regional habitat: Dune forests from Washington to California.

Comments: The maggots live inside blotch mines on the leaves. Pupation occurs within the mine.

SNOW PLANT

Snow plant is such a strange plant that it immediately attracts attention. It has no chlorophyll so how does it survive? Scientists are discovering today that many plants, including snow plant, demonstrate what is known as mycoheterotrophy, a condition where the plant parasitizes a mycorrhizal fungus associated with a neighboring plant root. This is how snow plant obtains its nourishment. The fine details of this association have yet to be elucidated. Snow plant's placement in the same family as coast huckleberry is obvious when one carefully examines the blooms. The little five-tipped, bell-shaped flowers closely resemble those of other members of the Ericaceae.

SNOW PLANT
(*Sarcodes sanguinea*: Ericaceae)
Characteristics: Stalk stout, fleshy, mostly reddish; lacking chlorophyll. Leaves scale-like. Flowers closed, five-lobed, bell-shaped.
Regional habitat: Dune forests in Oregon and California.

Comments: This parasitic perennial, which is related to and has flowers shaped like those of the coast huckleberry, depends on mycorrhizal fungi for its nourishment.

PINEDROPS

Pinedrops is another curious plant in dune forests that lacks chlorophyll. The plants look like slender spears that have been thrust up from beneath the ground. The flowers resemble those of the huckleberry and the mature seeds are slightly winged like those of pine, which is how the plant received its name. Since pinedrops lacks chlorophyll, it obtains nourishment in the same manner as snow plant, by parasitizing mycorrhizal fungi, in particular, members of the genus *Rhizopogon*, whose fruiting bodies (sporocarps) are known as "false truffles" or "potato mushrooms" because of their shape. *Rhizopogon* is a widespread genus and a member of the Shore Pine (*Pinus contorta*) Community along Pacific Northwest sand dunes (Poinar 2016).

PINEDROPS
(*Pterospora andromedea*: Ericaceae)

Characteristics: Stem thick, straight, reddish-brown, fleshy and sticky when young but drying with age. Leaves tiny, lance-like, clasping the stem; only persisting at base. Flowers urn-shaped, fused; petals five-lobed, pink, white, or yellowish. Fruit a woody capsule with winged seeds.

Regional habitat: Dune forests along the Pacific Northwest.

Comments: Pinedrops is another perennial that parasitizes mycorrhizal fungi associated with conifers.

MUSHROOMS

The sudden appearance of mushrooms and other fungi in various coastal dune habitats is a reminder that subterranean life, about which we know very little, is an integral part of coastal ecology. Beneath the surface of the ground lives a variety of fungi, bacteria, algae, protozoa, and nematodes, all intertwined with one other. These hidden organisms influence the growth of dune plants as well as their invertebrate associates, which usually have at least one developmental stage under or on the surface of the ground. Of the different types of saprophytic, parasitic, or symbiotic soil fungi, mushrooms and their close relatives are the most familiar. Saprophytic forms, such as conks, are important decomposers of dead wood. Parasitic forms, such as wine-caps, attack other mushrooms. Symbiotic forms, such as boletes, chanterelles, and earthballs, construct mycorrhizal sheaths around conifer rootlets and absorb nutrients and moisture. In return, they supply the conifers with essential minerals and organic compounds. This is why spores of symbiotic fungi are inoculated on conifer seedlings in many nurseries.

One of my favorite "mushrooms" is the earthstar. Not only is its shape bizarre but also its behavior of curving the spreading "arms" inward against the spore sac as it dries, thus forcing out the spores, which are food for springtails and fungus gnats. Another favorite is the coral fungus, shaped like an intricate sculpture created by unknown creatures working throughout the night. Then there are the subterranean potato mushrooms and earthballs that hardly appear to be related to mushrooms at all.

The candy and cream mushroom is quite spectacular when it first forms, with red droplets spread over a white cap. For interesting designs, there are morels with their caps resembling deformed brains. For bold hues, there are not many mushrooms as brilliantly colored as the red fly agaric. These various dune forest fungi and their insect associates represent a little studied segment of coastal dune life. Mention should be made that mushrooms are also appreciated as food by forest animals such as deer and squirrels.

MUSHROOM COMMUNITY **Herbivores**
Fungus gnat
Scavenger beetle
Phoridae
Springtails
Sap beetles
Mushroom flies
Spider fungus
Fly fungus

FLY AGARIC

(*Amanita muscaria*: Amanitaceae)

Characteristics: Cap red to orange-red with white patches (remains of veil). Stalk white, with a veil and bulbous base. Gills white, free, or slightly attached.

Regional habitat: In and along edges of dune forests in the Pacific Northwest.

Comments: Fungus gnat maggots (Diptera: Mycetophilidae) feed on the cap. Fly agaric was so named for producing secretions that are attractive, but also toxic, to some flies.

KING BOLETE

(*Boletus edulis*: Boletaceae)

Characteristics: An early stage of this mushroom resembles a bun over a layer of cream cheese. The mature stage has a reddish-brown convex cap with numerous pores on the undersurface and a firm, thick, reddish-brown stalk with a white bulbous base.

Regional habitat: Found near conifers along the Pacific coast.

Comments: King bolete forms a fungal sheath around the rootlets of conifers and assists in the absorption of minerals while the conifers provide moisture and organic compounds to the fungus. A scavenger beetle (*Necrophilus hydrophiloides*: Coleoptera: Agyrtidae) feeds on the stalk of king bolete throughout the Pacific Northwest coast.

SLIPPERY JACK

(*Suillus umbonatus*: Suillaceae)

Characteristics: Cap dingy tan; convex then becoming plane with margin uplifting. Stalk whitish; veil semi-gelatinous.

Regional habitat: Usually gregarious under conifers in the Pacific Northwest.

Comments: This species is not considered edible, possibly due to its slimy covering.

CHANTERELLE
(*Cantharellus* sp.: Cantharellaceae)
Characteristics: Cap yellowish-brown, flat, depressed or vase-shaped, with wavy margin. Gills and stalk similar in color to cap.

Regional habitat: Along the Pacific Northwest coast.
Comments: This species is often gregarious and occurs with conifers such as Douglas-fir and Sitka spruce. It is said to be edible.

PACIFIC GOLDEN CHANTERELLE
(*Cantharellus formosus*: Cantharellaceae)
Characteristics: Cap yellowish-tan with reflexed margins. Stalks thick.
Regional habitat: Appears singly or in groups in conifer forests along the Pacific Northwest coast.
Comments: This common edible species is Oregon's State Mushroom.

COFFEE CUP

(*Coltricia* sp.: Hymenochaetaceae)

Characteristics: Cap small, circular, fibrous; dark brown with cream-colored border; underside with yellowish-brown pores. Stalk reddish-brown, fairly thick, tough.

Regional habitat: Along the Pacific Northwest coast.

Comments: Coffee cup is a stalked polypore and grows singly among moss and fallen leaves. From the top it looks like a cup of coffee.

FAIRY STOOL

(*Coltricia perennis*: Hymenochaetaceae)

Characteristics: Cap reddish-brown with irregular border and circular zones of different hues. Stalk brownish, thick, central.

Regional habitat: On ground in forests along the Pacific Northwest coast.

Comments: Fairy stool is widespread the year around, but not edible.

CORAL FUNGUS

(*Ramaria* sp.: Gomphaceae)

Characteristics: Fruiting structure consists of a mass of branching short, fleshy, pinkish stalks bearing spores on the surface.

Regional habitat: Associated with conifers along the Pacific Northwest.

Comments: This beautiful fungus, that resembles corals, also can be yellow, red, or orange.

UMBRELLA CAP MUSHROOM

(*Enteloma* sp.: Entolomataceae)

Characteristics: Cap wide, tan; center convex, often with slight depression in raised portion. Gills attached to short, narrow stalk.

Regional habitat: Widespread along the Pacific Northwest coast and elsewhere.

Comments: This mushroom also occurs in dunes quite close to the ocean.

ELFIN SADDLE

(*Helvella* sp.: Helvellaceae)

Characteristics: Cap folded, wrinkled, irregular, dark blue to black. Cap erect or bent at an angle from the top of the fluted, erect stem.

Regional habitat: Forest clearings and along trails in Oregon

Comments: This ascomycete is associated with conifers.

STRAWBERRIES AND CREAM

(*Hydnellum peckii*: Bankeraceae)

Characteristics: Cap white, wide, somewhat uneven and flat, covered with various-sized red drops. Stalk thick, bulbous with brown spiny hairs.

Regional habitat: Widespread along the edges of Pacific Northwest dune forests.

Comments: While this beautiful mushroom looks appetizing, the flesh is somewhat spicy. The red drops on the cap are said to have anticoagulant properties.

HEDGEHOG MUSHROOM
(*Sarcodon imbricatus*: Bankeraceae)
Characteristics: Cap pinkish, wide, smooth. Fruiting body under cap composed of tiny spines. Stalk thick, white, bulbous.
Regional habitat: Dune conifer forests, especially around Sitka spruce, throughout the Pacific Northwest.
Comments: Hedgehog mushroom resembles a gilled mushroom from the top. It is supposedly edible but somewhat bitter.

GRAY MYCENA

(*Mycena* sp.: Mycenaceae)

Characteristics: Cap gray, first conical, then expanding with age to almost flat, becoming thin enough to see gills beneath; often covered with white remains of veil. Stalk thin, fragile.

Regional habitat: Gregarious under conifers in and along the edges of dune forests.

Comments: Mycenas are very small and fragile mushrooms that are usually not considered edible.

EARTHSTAR

(*Astraeus* spp.: Diplocystaceae) and (*Geastrum* spp.: Geastraceae)

Characteristics: Outer tissue thick, splits into star-like rays. Center spore case initially appears like a small puffball, ruptures irregularly or spores are released by a single apical pore.

Regional habitat: These are cosmopolitan ectomycorrhizal fungi associated with a range of conifer and deciduous trees.

Comments: Earthstars are not considered edible for humans, however, many insects, especially springtails and fungus gnat larvae, feed on the spores (note fungus gnat larvae in opened spore case).

DELICIOUS MILK-CAP

(*Lactarius deliciosus*: Russulaceae)

Characteristics: Medium-sized fleshy, yellowish mushroom with decurrent gills; veil absent; cap depressed, surface somewhat viscid. Bruised areas turn greenish.

Regional habitat: From Washington to California.

Comments: This is an ectomycorrhizal species that is associated with conifers.

COMMON MOREL
(*Morchella esculenta*: Morchellaceae)
Characteristics: Cap light brown with strong vertical ribs. Stalk short, light tan to almost white, attached to base of cap.

Regional habitat: Widespread along the Pacific Northwest coast.
Comments: This species also occurs in sand dunes, apparently obtaining nourishment from buried logs.

VELVET-TOP

(*Phaeolus schweinitzii*: Fomitopsidaceae)

Characteristics: Caps one to many overlapping; rusty-brown, semicircular, tough with woolly surfaces and wavy margins. Pores greenish-yellow. Stalk absent or very short and thick. If several plants occur together, their stalks may be fused.

Regional habitat: In conifer forests along the Pacific Northwest coast.

Comments: Velvet-top is a parasite and can be pathogenic to conifers.

SHAGGY MANE
(*Coprinus comatus*: Agaricaceae)
Characteristics: Cap cylindrical, large, white, covered with large partially exfoliating tan-tinted scales. Stalk slender, straight, white.

Regional habitat: Often found in groups along the edges of dune forests.
Comments: Considered an excellent edible species. Individuals can reach over a foot in height.

WHITE MARASMIUS
(*Marasmiellus* sp.: Marasmiaceae)
Characteristics: Small, white mushroom with a delicate, pleated cap, widely spaced gills and short, curved stalk.

Regional habitat: Widespread along the Pacific Northwest coast.
Comments: This species grows on twigs and stems of various plants, especially shrubs.

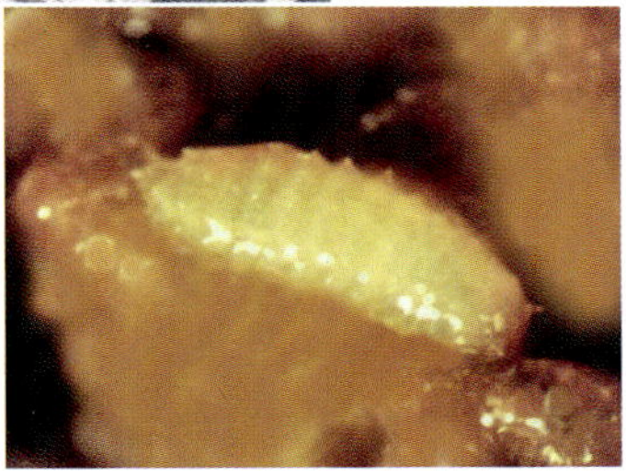

EARTHBALL

(*Scleroderma* sp.: Sclerodermataceae)

Characteristics: Roundish spore case with hard outer wall (peridium) that splits open to release a mass of sooty spores.

Regional habitat: Dune forests along the Pacific Northwest coast.

Comments: Earthball obtains nourishment from the roots of various conifers. The hard outer wall of earthballs distinguishes them from puffballs. Sap beetle larvae (Coleoptera: Nitidulidae) can be found developing inside the spore cases. They have short, straight, yellowish bodies, small legs, and a pair of spurs at the tail end.

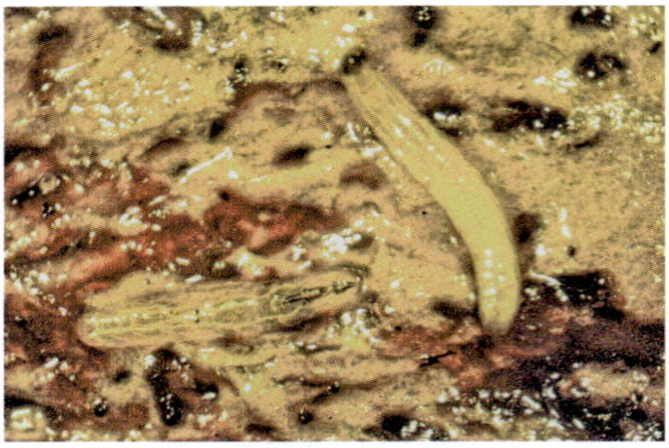

MUSHROOM FLIES

(*Diptera*: Sciaridae and Phoridae)

Characteristics: Sciarid fungal gnats (vertical larva on right in photo) have a uniformly pale body and solid black head. Phorid fungal maggots (horizontal maggot on left in photo) have a partially transparent body and small black mouth hooks.

Regional habitat: Pacific Northwest.

Comments: Many insects feed on mushrooms, but flies predominate. These are two of the most common fly families that feed on mushrooms.

SPIDER FUNGUS

(*Gibellula* sp.: Cordycipitaceae)

Characteristics: Develops inside arachnid, then produces elongate external conidiophores bearing asexual spores. The sexual stage of this genus is a species of *Torrubiella*.

Regional habitat: Pacific Northwest.

Comments: This genus appears to be specific on spiders. All known species eventually kill their hosts.

FLY FUNGUS

(*Aspergillus* sp.: Trichocomaceae)

Characteristics: Mycelium develops inside fly while sporulation occurs on surface of host. Conidia are white and borne in chains at the tips of conidiophores.

Regional habitat: Pacific Northwest.

Comments: Members of this ubiquitous genus can function as saprophytes, parasites, or pathogens of invertebrates and vertebrates. Several species are known insect pathogens.

DUNE FOREST VERTEBRATES

A number of vertebrates occur in Pacific Northwest dune forests and some of the more obvious ones that hikers are likely to see as they explore plant communities are presented below. Some invertebrates that are associated with these vertebrates are also included.

Vertebrates are an important component of both dune forest and dune wetland plant communities. By feeding on plants, distributing seeds, preying on herbivorous insects and nutrient recycling, they maintain a state of natural balance with the environment.

ELK

(*Cervus canadensis*: Cervidae)

Characteristics: Large, tan to light-brown with yellowish-white rump patch. Males with mane and large spreading antlers.

Regional habitat: Edges of dune forests from Washington to northern California.

Comments: Elk feed on a variety of plants. They usually occur in large groups consisting of a male, that makes a bugling call, and several females with young.

MULE DEER

(*Odocoileus hemionus*: Cervidae)

Characteristics: Body brown; often reddish in summer; ears large; tail black-tipped.

Regional habitat: In and along edges of dune coniferous forests throughout the Pacific Northwest.

Comments: There is some confusion regarding the identification of mule, black-tailed and white-tailed deer in North America. Since their ranges overlap, interbreeding occurs and there are a number of varieties, it is possible that they represent a single species.

DEER KED

(*Lipoptena* sp.: Diptera: Hippoboscidae)

Characteristics: Small, brown, flattened fly with a wide head, large eyes and sturdy legs. The large fragile wings extend well behind the body.

Regional habitat: Parasitic on deer throughout the Pacific Northwest.

Comments: This species differs from the introduced Old World deer keg (*Lepoptena cervi*) that occurs on white-tail deer in eastern North America. All larval development occurs in the body of the female deer ked. She then deposits a white pre-pupa that darkens on the forest floor. Winged adults emerging from the pupae search for deer.

BLACKLEGGED TICK

(*Ixodes pacificus*: Ixodidae)

Characteristics: Females and female and male nymphs reddish-brown to white with longitudinal black bars. Males black with white rim around edge of body. Scutum black in both sexes.

Regional habitat: Adult blacklegged ticks feed on deer (this species is also called the deer tick) throughout the Pacific Northwest.

Comments: Larvae prefer to feed on mice. Aside from deer, the nymphs and adults also feed on rodents, cats, dogs, and humans. Both nymphs and adults can transmit Lyme disease, babesiosis, and anaplasmosis to mammals, including humans.

BLACK BEAR

(*Ursus americanus*: Ursidae)

Characteristics: Body black (black phase), without shoulder hump; face black to brown; claws short.

Regional habitat: Forested areas along the Pacific Northwest.

Comments: Black bears are primarily nocturnal but will forage during the day when hungry. They are usually solitary except for a mother with cubs. Eating habits are variable and food includes fruits, small animals, and garbage. Dens are often made in the ground beneath trees.

RACCOON

(*Procyon lotor*: Procyonidae)

Characteristics: Size of a small dog; body covered with grizzled hair; claws non-retractile; black mask over eyes; tail with alternating light and black bands.

Regional habitat: Dune forests and wetlands from Washington to California.

Comments: Raccoons are usually nocturnal but will forage during the day when hungry. They are solitary except for a mother with young. They are omnivorous and prefer foraging along the borders of dune lakes and ponds for food. Dens are often made in trees.

OPOSSUM

(*Didelphis virginiana*: Didelphidae)

Characteristics: Body gray with some longer, scattered white hairs; head white, pointed; ears black with white tips; tail rat-like, long, white with gray base; snout and tips of digits pink.

Regional habitat: Forested areas along the Pacific Northwest.

Comments: Opossums are slow-moving and nocturnal, eating a variety of plant and animal matter. This is the only marsupial in North America.

DEER MOUSE

(*Peromyscus maniculatus*: Cricetidae)

Characteristics: Body yellowish-brown to gray; tail bi-colored (dark above, white below), usually as long as or longer than head and body combined.

Regional habitat: Throughout the Pacific Northwest.

Comments: Deer mice eat seeds, fruits, and various invertebrates. They nest in burrows or tree stumps. They are attacked by a number of predators, infested by ticks and mites, and parasitized internally by bot flies and helminth worms.

BOT FLY

(*Cuterebra* sp.; Diptera: Oestridae)

Characteristics: Mature maggot reddish-brown, oval to round; pupates immediately after leaving host. Adult large, robust, grayish; thorax with white patch on both sides; eyes opaque, widely separated, slightly protruding; wings brownish-gray.

Regional habitat: Throughout the Pacific Northwest.

Comments: The larvae are internal parasites of rats, mice, and rabbits.

TROWBRIDGE SHREW

(*Sorex trowbridgei*: Soricidae)

Characteristics: Brownish body with lighter under parts; tail bi-colored (brown above, white below).

Regional habitat: Conifer forests along the Pacific Northwest.

Comments: While shrews are predators of insects and other invertebrates, they have been reported to eat conifer seeds.

HORSEFLY

(*Tabanus* sp.: Diptera: Tabanidae)

Characteristics: Large, robust, brown fly; abdomen with faint black medial longitudinal stripe; head bearing large metallic eyes; antennae short, straight or slightly curved; wings mostly clear with some tinting along upper margin.

Regional habitat: Pacific Northwest.

Comments: Females of this genus bite a number of animals, including humans. The bite is painful and can be serious since horseflies can transmit pathogens such as tularemia and anthrax.

DEERFLY

(*Chrysops* sp.: Diptera: Tabanidae)

Characteristics: Medium-sized fly with black transverse bands on abdomen; each wing with a broad black median bar.

Regional habitat: Pacific Northwest.

Comments: Deerflies attack a number of animals, including humans. Their bite is painful and can transmit anthrax.

ROUGH-SKINNED NEWT

(*Taricha granulosa*: Salamandridae)

Characteristics: Brown above, yellow to reddish below; skin rough, pitted; eyes small; tail laterally flattened.

Regional habitat: Forests and wetlands from Washington to California.

Comments: This terrestrial phase of the rough-skinned newt occurs under bark or moss, often far from water. It feeds on earthworms and small arthropods. In the aquatic breeding phase, the skin becomes smooth and the tail more flattened. Rough-skinned newts are poisonous and should be handled with care.

NORTHWESTERN SALAMANDER
(*Ambystoma gracile*: Ambystomatidae)
Characteristics: Shiny, grayish body with a few scattered white flecks; head broad; parotid glands in neck region enlarged (unless secretions recently released against a predator).
Regional habitat: Forests along the Pacific Northwest.
Comments: White deposits on back of the head are partially dried secretions released from parotid glands.

WESTERN SKINK

(*Plestiodon skiltonianus*: Scincidae)

Characteristics: Small, elongate lizard with smooth scales displaying wide color variations in different populations. Oregon populations with tan back, two brown lateral stripes running from nose to groin and dark brown legs.

Regional habitat: Throughout the Pacific Northwest.

Comments: Western skinks are diurnal and prey on insects and spiders. When startled, a skink is capable of burrowing in the sand as well as running rapidly. The two to six eggs deposited under debris are attended by the female.

NORTHERN ALLIGATOR LIZARD
(*Elgaria coerulea*: Anguidae)
Characteristics: Large, brown body with a grayish head; broken vertical rows of white scales on sides; eyes yellow with black pupils.

Regional habitat: Throughout the Pacific Northwest.
Comments: These lizards are quite agile and occur in sand dunes near the ocean as well as in coniferous forests.

PACIFIC GOPHER SNAKE
(*Pituophis catenifer*: Colubridae)
Characteristics: Basically yellowish with variously sized dark blotches covering dorsal and lateral surfaces.
Regional habitat: Open coniferous dune forests along the Pacific coast.

Comments: This non-venomous snake is active by day and kills prey (other vertebrates) by constriction. It can simulate a rattler by coiling its body and vibrating its tail.

WESTERN GARTER SNAKE
(*Thamnophis elegans*: Colubridae)
Characteristics: Slender, brownish snake; a red or yellow mid-dorsal line and yellow lateral lines extend the length of the body.
Regional habitat: Along the Pacific coast from Washington to California.
Comments: Color patterns of western garter snakes vary considerably. When captured, garter snakes release feces and odiferous secretions from their anal glands. Garter snakes are non-venomous livebearers and consume a wide range of invertebrates and small vertebrates.

OSPREY

(*Pandion haliaetus*: Pandionidae)

Characteristics: Large bird with dark back and wings, white underparts and head with dark, broad lateral patches extending slightly beyond the eyes.

Regional habitat: Widespread in all forests along the Pacific coast.

Comments: It is fascinating to watch the osprey dive feet first in the water to catch fish in dune lakes. It is also amazing how the adults can keep their bulky nests balanced at the tips of dead trees.

TURKEY VULTURE

(*Cathartes aura*: Cathartidae)

Characteristics: Large, black to brown bird with a naked, red head (dark head in immatures); wingspan often over two meters.

Regional habitat: Throughout the Pacific Northwest.

Comments: These large birds make excellent use of wind currents and can soar in the sky for extended periods while looking for carrion. After a rainstorm, they can be seen resting on tree branches or fallen logs drying their outstretched wings.

AMERICAN GOLDFINCH
(*Carduelis tristis*: Fringillidae)
Characteristics: Small songbird. Male bright yellow with black forehead and wings; female dull yellow without black forehead.

Regional habitat: Woodlands and other habitats along the Pacific coast.
Comments: Food is mainly seeds with insects as a side dish.

SPOTTED TOWHEE
(*Pipilo maculatus*: Passerellidae)
Characteristics: Large songbird. Bill short, stout; tail patches white, obvious during flight. Male with black head and chest and white belly with lateral rufous patches. Female similar but with brown head and chest.

Regional habitat: Forests and other varied habitats along the Pacific coast.
Comments: Adults search for food under leaves. Males often bring food to the females while they are incubating the eggs.

STELLER'S JAY
(*Cyanocitta stelleri*: Corvidae)
Characteristics: Large, blue bird with a dusky black head and neck; top of head with long black crest; voice a loud harsh "caaa."

Regional habitat: Conifer forests along the Pacific Northwest coast.
Comments: Steller's jays are quite intelligent as well as bold, even entering camp tents to obtain food.

CEDAR WAXWING

(*Bombycilla cedrorum*: Bombycillidae)

Characteristics: Medium-sized bird with white head and breast; head with short crest and black triangular mask covering eyes; tail with terminal yellow band.

Regional habitat: Forests along the Pacific Northwest coast.

Comments: The black mask and head crest quickly identifies the cedar waxwing.

FOX SPARROW
(*Passerella iliaca*: Passerellidae)
Characteristics: Small, brownish bird with white chin, finely streaked breast and dusky area at back of neck.

Regional habitat: Forests along the Pacific Northwest coast.
Comments: This is a rather timid sparrow and difficult to observe.

WHITE-CROWNED SPARROW
(*Zonotrichia leucophrys*: Passerellidae)
Characteristics: Small, white-breasted bird with a black-and-white striped head and yellow bill.
Regional habitat: Forest borders and other habitats along the Pacific Northwest coast.
Comments: This sparrow is quite bold and is an expert at finding cutworms and other insects in plant debris.

RAVEN

(*Corvus corax*: Corvidae)

Characteristics: Large, solid black bird with a stout, triangular, black bill and white neck feathers (best seen when wind ruffles the feathers).

Regional habitat: Forests and various other habitats along the Pacific Northwest coast.

Comments: Often solitary, the raven is best seen near wet areas around forest clearings.

PILEATED WOODPECKER
(*Dryocopus pileatus*: Picidae)
Characteristics: Large, black bird with a bright red crest; males with red stripe extending from bill to base of jaw.
Regional habitat: In conifer or mixed forests along the Pacific Northwest.
Comments: This is one of the largest woodpeckers in North America.

RINGED TURTLE DOVE
(*Streptopelia risoria*: Columbidae)
Characteristics: White bird with a grayish-blue bill, partial black ring around the back of the neck and pink feet.

Regional habitat: Occurs sporadically in dune forests of the Pacific Northwest.
Comments: This is an escapee of the domestic African turtledove. It has now become adjusted to the cool, wet conditions of the Pacific Northwest.

WILD TURKEY
(*Meleagris gallopavo*: Phasianidae)
Characteristics: Large bird that resembles the domestic turkey; head naked, bluish with red neck lobe (wattle); tail feathers with white tips.
Regional habitat: Found along borders, in clearings and within dune forests along the Pacific Northwest coast.
Comments: In dune forests, they feed on berries of salal and coast huckleberry. Wild turkeys can fly and often spend the night in trees. They also have a temper and can be dangerous. Some California populations are infected with Lyme disease.

INTRODUCTION TO DUNE WETLAND PLANT COMMUNITIES

Dune wetlands are unique habitats that offer many opportunities for study. They consist of various aquatic domains that occur near the coast to deep in the forests. Some small saline ponds form quite close to the sea and while the water is slightly salty, it supports the development of tadpoles and aquatic insects. Freshwater wetlands occur further inland and range from marshes, bogs, and rivulets to ponds and large lakes. These usually are found within the boundary of dune forests, however they may be exposed when the surrounding vegetation has been removed. Some of these inland water sources are temporary, forming during the winter rains and drying out in the summer. Others are fed by springs or ground water and have existed for many years. Many of the smaller dune forest lakes that sustain beaver populations and other aquatic life are slowly disappearing. Some of the larger dune lakes, like Siltcoos Lake and Devil's Lake in Oregon and Lake Earl in California offer hiking, birding, fishing, boating, and other recreational facilities. While pleasant for tourists, the associated construction affects the natural vegetation and introduced fish and exotic water plants upset the native aquatic life.

TEMPORARY COASTAL POND

This pond fills with rainwater during heavy storms and contains a range of aquatic life, being a good source of tadpoles in the spring. Tinted with salt, these coastal ponds offer a convenient habitat to examine fresh-water organisms that have adapted or are adapting to a slightly saline aquatic environment.

TEMPORARY INLAND POND

This freshwater pond is about half its full size and, as summer progresses, will disappear completely. Such temporary ponds contain a variety of aquatic life and are much appreciated by great blue herons that feed on the tadpoles.

SMALL DUNE LAKE

This dune lake is surrounded by spruce and hemlock. Beaver, as well as various wading birds, can be seen along the shoreline.

LARGE DUNE LAKE

This dune lake is one of the largest not affected yet by tourism. Osprey nests cap some of the tall conifer trees and minnows skirt the banks. It is the perfect place for meditation.

SLOUGH SEDGE

Slough sedge, which occurs along the edges of many dune bogs and marshes, produces two separate flower clusters. The lower female flowers have thin, white styles while the upper male flowers support rows of yellowish stamens. Small adult rove beetles arrive to feed on maturing pollen, which is unusual since these beetles normally are predators.

SLOUGH SEDGE COMMUNITY	
Herbivores	**Commensal**
Rove beetle	Mites
Aquatic leaf beetle	

SLOUGH SEDGE

(*Carex obnupta*: Cyperaceae)

Characteristics: Perennial. Leaves tall, stiff, flat; male and female flowers borne separately on longish spikes. Female flowers with white styles occur beneath male flowers bearing yellowish stamens.

Regional habitat: Dune wetlands from Washington to California.

Comments: Rove beetles feed on the pollen. Leaves of slough sedge are still used in basket-making.

ROVE BEETLE

(*Eusphalerum* sp.: Coleoptera: Staphylinidae)

Characteristics: Small, brown beetle with ocelli and a large rectangular pronotum with slightly rounded margins; elytra are abbreviated (only reach middle of abdomen).

Regional habitat: The genus is widespread along the Pacific coast.

Comments: The adults are obligate pollen-feeders. Many carry phoretic mites under their wing covers. The larvae probably develop in litter at the base of slough sedge.

AQUATIC LEAF BEETLE

(*Plateumaris* sp.: Coleoptera: Chrysomelidae)

Characteristics: Large metallic brown to black beetle with a narrow pronotum; elytra with rows of small scales.

Regional habitat: On slough sedges in wetlands of the Pacific Northwest.

Comments: The adults visit slough sedge for pollen. The larvae are aquatic and obtain air by inserting the tips of their tube-like abdomens into hollow stems of aquatic plants.

MONTEREY CENTUARY

This native annual-biennial, with its wide-open, expanded, pinkish to white petals, prefers to grow in moist meadows at the borders of dune forests along the Pacific Northwest. While small, the flowers stand out clearly against the green background. However, I never saw any herbivores on the plant or bees coming for nectar. Perhaps this is due to some chemicals in the plant cells since a bitter tonic made from dried plants of Monterey centuary is taken to increase appetite and enhance stomach functions.

MONTEREY CENTUARY
(*Zeltnera muehlenbergii*: Gentianaceae)
Characteristics: Small annual or biennial. Leaves opposite, narrow. Flowers with five white to pink petals; borne in terminal clusters.

Regional habitat: Most often in low, damp habitats, especially in southern Oregon and California.
Comments: This plant is found in the same habitat as slough sedge.

WATER HEMLOCK

Along the borders of dune lakes and other dune wetlands, with their stems often covered by water, are tall plants of water hemlock bearing clusters of intricate, white flowers. The miniature flowers are quite striking in their symmetry of the 20 to 30 blooms. Yet this innocent-appearing plant is considered one of the most toxic plants in North America. The poison is a cicutoxin that occurs throughout the plant tissues, with the highest concentrations in roots and rootstalks. Apparently, the dose is lower in the green leaves and fruits, which is why deer ingest these parts unharmed. Its close relative, poison hemlock, with the same group of toxins, is famous for being the plant that killed Socrates. The symptoms caused by cicutoxin can be quite painful, often resulting in seizures and respiratory failure that may occur as quickly as 15 minutes after consumption. Poison in the nectar and pollen must be absent or have no effect on insects, since the pollinating sand wasps do not appear to be affected.

WATER HEMLOCK COMMUNITY Herbivores
Sand wasp
Bees

WATER HEMLOCK
(*Cicuta douglasii*: Apiaceae)
Characteristics: Tall perennial. Stems hollow, erect, from taproot. Leaves basal, divided one to three times; borne along stem. Flowers small, white to greenish, in terminal or lateral clusters.
Regional habitat: Marshes, edges of streams, ponds and lakes from Washington to California.

Comments: Various insects, including bees and sand wasps (see specimen in photo), visit the flowers. The poison in water hemlock can be lethal to humans and livestock if any portion of the plant is eaten. The roots may be mistaken for wild parsnips. However, skin irritation from touching the plant is rare.

DODDER

With yellow, thin, wiry stems that curl around the host plant, dodder is one of the strangest plants I have encountered in wetland habitats. Other common names, like devil's hair, hellbine, and strangleweed, reflect human sentiments of this plant. With reduced levels of chlorophyll, dodder's survival depends on parasitizing green plants. Imagine scenes from a horror movie depicting the coils of a serpentine vine twisting around your limbs, then sending ducts through your skin to retrieve your blood. Plants parasitized by dodder undergo a similar scenario when they become helplessly entwined and pierced. To add insult to injury, dodder passes pathogenic viruses into the plants it is parasitizing. Many of these viruses are symptomless in dodder but can ravage their hosts, causing mottling and necrosis of leaves, stunting of stems, and tumors. The only part of the dodder plant that can be considered "normal" is the five-part, tiny, white to yellow flower that arises directly from the stem. It is curious that the seeds of dodder are said to have health benefits and are claimed to contain anti-aging compounds.

DODDER

(*Cuscuta salina*: Convolvulaceae)

Characteristics: Annual parasitic vine. Stems orange-yellow, capable of twisting around host plants. Leaves reduced to minute swellings on stems. Flowers small, white, with five petals; borne in axillary clusters.

Regional habitat: Salty marshes, ponds, and bogs from Washington to California.

Comments: After wrapping itself around a plant stem, dodder produces small suckers (haustoria) that penetrate into the tissues of the host and withdraw water and nutrients. Once established, dodder's root system dries up. Various host plants are attacked in wetland habitats, especially members of the amaranth (Amaranthaceae) and sunflower (Asteraceae) families.

COBRA PLANT

Strange, stiffly upright leaves with bent-over "heads" give the cobra plant its name. Another name is pitcher plant, referring to the enclosed area formed by the hollow leaves. It must be odors coming from the tiny entrance holes in the curved heads that are so attractive to insects. Passing through the entrance hole and moving down the slippery reflexed hairs lining the passageway is easy enough; getting back out is the snag. Moving back up the leaf over downward pointing hairs is an impossible task for most insects. The twisted top of the pitcher plant now hides the small entrance hole and attempts to exit soon tire the victim. Eventually, the insect starts sliding downward, landing in a watery solution at the bottom of the leaf, a solution filled with enzymes and bacteria. Its disintegrating body will then join the remains of other creatures that were also tricked by the clever cobra plant.

What an ingenious method to obtain nourishment. The tubular cobra plant leaves I opened were filled with insect cadavers. But, surprisingly, among the deteriorating bodies were some developing fly maggots. Over the eons, some insects evolved a body coating that protected them from the enzymes and allowed them to dine on the plant's unfortunate victims. Thus, while insects ranging from small flies to large wasps undergo digestion in the pitchers, midges, fruit flies, and scuttle flies freely scavenge their decomposing bodies in this complex microcosm.

COBRA PLANT COMMUNITY Detritivores
Small fruit fly
Midge
Scuttle fly
Mite
Nematode
Ciliate

COBRA PLANT

(*Darlingtonia californica*: Sarraceniaceae)

Characteristics: Insectivorous perennial with tall, erect stems. Leaves (pitchers) yellowish to greenish, red-dotted; vase-shaped with reflexed, hood-like top. Flowers with five greenish, overlapping sepals enclosing five purple, bell-shaped petals; borne at the tips of long stalks.

Regional habitat: Bogs and marshes in California and Oregon.

Comments: The inner surfaces of the leaves are lined with reflexed, slippery hairs to keep trapped insects from exiting. The digested cadavers, including this yellowjacket, release nutrients utilized by the plant.

SMALL FRUIT FLY

(Diptera: Drosophilidae)

Characteristics: Larva white; legless. Adult small, brown; eyes large, red; wings fuscous.

Regional habitat: In cobra plants in Oregon and California.

Comments: The larvae feed on decaying insects and associated microorganisms. They appear to be immune to the cobra plant's digestive enzymes.

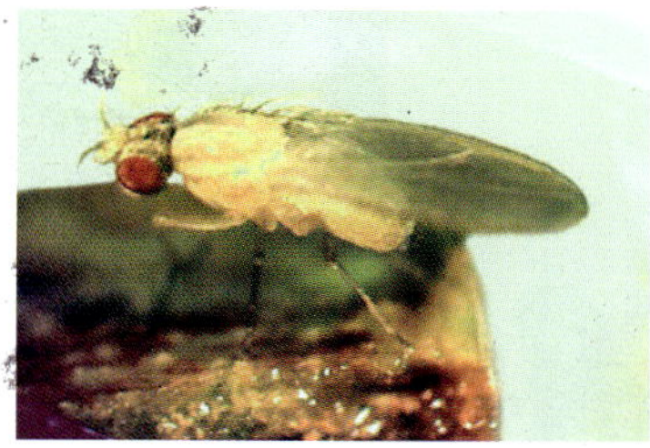

NEMATODE

(Nematoda: Rhabditoidea)

Characteristics: Small, elongate, worm-like; segments and appendages lacking.

Regional habitat: In cobra plants in Oregon and California.

Comments: All developmental stages feed on microorganisms associated with the decaying insects in the pitchers of the cobra plant. Third-stage juveniles (dauer) stand with their tail tips on setae lining the inner sides of the pitcher. In this position, they can attach themselves to flies leaving the pitchers.

MIDGE

(*Metriocnemus edwardsi*: Diptera: Chironomidae)

Characteristics: Eggs black, elliptical. Maggots yellowish, elongate, legless with brown head. Pupae yellowish, with pointed tail and reddish-brown eyes. Adults small, brownish-gray to black; male with feathery antennae.

Regional habitat: Found only in cobra plants in Oregon and California.

Comments: This is the most common insect developing in pitchers of cobra plants.

SCUTTLE FLY

(*Megaselia* sp.: Diptera: Phoridae)

Characteristics: Maggots small, white with short projections on body margins; head pointed; legless. Adults small, black, with dark eyes, short antennae and wings held flat against back at rest.

Regional habitat: In cobra plants in Oregon and California.

Comments: All stages occur in pitchers of cobra plants.

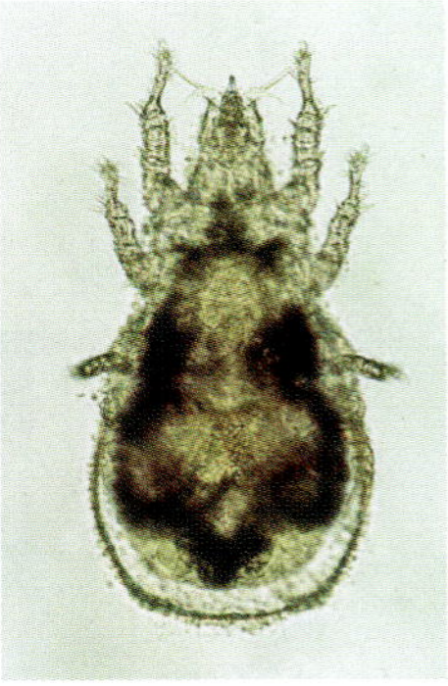

MITE

(*Sarraceniopus darlingtoniae*: Astigmata: Histiostomatidae)

Characteristics: Drop-shaped mite with pointed head; legs six (larva) or eight (adult).

Regional habitat: Only in pitchers of cobra plants in Oregon and California.

Comments: These mites are filter feeders and consume bacteria and protozoa that proliferate on the bodies of insects in cobra plants.

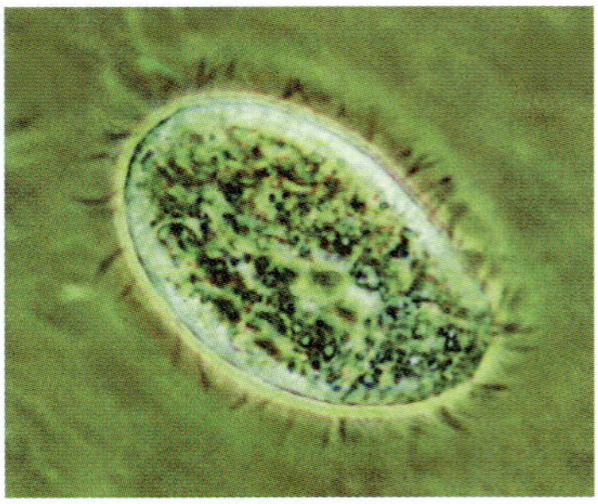

CILIATE

(Protozoa: Ciliophora)

Characteristics: Microscopic, oval, covered with short cilia (hairs).

Regional habitat: Only in pitchers of cobra plants in Oregon and California.

Comments: These ciliates assist in the breakdown of insect cadavers, thus making nutrients available to the cobra plant.

ROUND-LEAVED SUNDEW

Another insectivorous bog and marsh plant uses a completely different tactic to entrap insects. The little round-leaved sundew is, in many ways, the opposite of the towering cobra plant. With the sundew, there is no small hole for entering a trap. The receptacle is completely exposed and, instead of fetid odors to entice its victims, the sundew has leaves covered with stalks bearing sweet, viscous drops just waiting to be tasted. But that is the deception. Anything that touches these enticing drops becomes stuck. The stalks are capable of slowly bending inward on an entrapped victim, bringing it into contact with the leaf tissue. Eventually, the insect is digested, with only some chitinous parts remaining.

However, some very small, delicate insects elicit no response from the sundew and seem to know how to avoid making contact with the sticky drops on the leaves. Thus, tiny aphids and mites roam over the surface of the leaves with impunity, alongside the decomposing bodies of the less fortunate.

ROUND-LEAVED SUNDEW COMMUNITY Herbivores
Aphid
Mite

ROUND-LEAVED SUNDEW

(*Drosera rotundifolia*: Droseraceae)

Characteristics: Small perennial. Leaves viscous, glandular, bordered with long-stalked hairs that exude sticky drops. Flowers small, white; borne on erect stalks.

Regional habitat: Bogs, fens, and seeps throughout the Pacific Northwest. Often in association with sphagnum mosses, sedges, and grasses.

Comments: When an insect lands on an open leaf, the gland-tipped hairs hold it and the leaf folds up around the victim and slowly digests it.

APHID

(*Aphis* sp.: Hemiptera: Aphididae)

Characteristics: Small, grayish-green; cornicles (tubes at posterior end) extended; antennae long.

Regional habitat: Associated with round-leaved sundew in Oregon and California.

Comments: Aphids feed on the leaves of round-leaved sundew. Mites also roam over the leaves with impunity. It is unknown, but fascinating, how aphids and mites escape the glandular hairs that entrap other small insects.

YELLOW FLAG

Native to Europe, parts of Asia, the Middle East, and northern Africa, this tall iris, with its showy yellow flowers, dots the edges of dune lakes and other wetlands along the Pacific Northwest coast. It also has similar habitat preferences throughout many parts of North America. In these wet habitats, where the base of yellow iris is often well beneath the water surface, it prefers full sun and blooms in late spring to early summer. Its status as an invasive weed comes from its habit of forming large colonies that can displace native plants. Yellow iris has been used to produce dyes and inks and also has some medicinal properties. The seeds from the large pods have even been used as a coffee substitute.

YELLOW FLAG
(*Iris pseudacorus*: Iridaceae)
Characteristics: Tall perennial. Leaves long, narrow, basal. Flowers large, yellow; borne at top of leafless stems.
Regional habitat: Open areas in dune wetlands in Washington and Oregon.

Comments: This beautiful iris, which was introduced from Europe, is now common in many aquatic habitats of North America.

LABRADOR TEA

Leathery entire leaves with bent over margins and clusters of white flowers with extended stamens identify this evergreen shrub. The long stamens may be related to certain types of pollinators. With its aromatic properties, dried leaves of Labrador tea are used to make a laxative drink that is also said to calm upset stomachs. Other medicinal properties of fresh tinctures of this plant include its use against scabies (caused by itch mites, chiggers, or harvest mites) and fungal skin lesions. Like so many other dune wetland plants, Labrador tea has established symbiotic associations with ericoid mycorrhiza.

LABRADOR TEA
(*Rhododendron groenlandicum*: Ericaceae)
Characteristics: Evergreen shrub. Leaves leathery, alternate, entire, margins turned under. Flowers white; borne in clusters at stem tips; stamens exserted, slightly longer than style.

Regional habitat: Bogs, marshes, and lakesides from Washington to California.
Comments: Some people use the spicy leaves to make tea, but apparently the taste varies depending on the plant's habitat.

BOG CLUB MOSS

Club mosses are small, primitive evergreen plants that reproduce by spores. They resemble mosses in size and shape but a closer look shows that they have pointed, needle-like leaves shaped like those of some conifers. Those living in Pacific Northwest dune forests are ground dwelling in moist areas while tropical species ascend trunks and occupy lower tree limbs. Being placed in the Phylum Lycopoda or Division Lycopodiophyta, they are often referred to as "lycopods." The spores are formed in sporangia located at the bases of the leaves. One curious feature of the spores is that they burn quickly and intensely with little heat. Apparently, this feature was useful in the past to produce light effects in theaters.

BOG CLUB MOSS
(*Lycopodiella inundata*: Lycopodiaceae)
Characteristics: Small perennial. Leaves tiny, numerous, scale-like, overlapping, covering stems. Spore-producing stems erect while rest of plant usually prostrate. Spores produced in dense clusters (sporangia) in leaf axils.

Regional habitat: Widespread in bogs and seeps throughout the Pacific Northwest.
Comments: Club mosses are primitive, spore-producing plants that were dominant millions of years ago. I could not find any associated insects.

BOG BEAN

If you could have visited coastal habitats from Alaska to California some 10,000 years ago in early summer, you would have been greeted by numerous, fuzzy white flowers of bog bean, a legume that grew in various bodies of shallow water. That was when the temperature was about 3–4 degrees C cooler than today. You can still find the thick walled seeds from these early plants buried in peat bogs along with artifacts made by Early Americans. Some of the seeds have sharp broken edges, suggesting that the early peoples knew about medicinal properties of bog bean, which was used to lower fever and cure stomach disorders, rheumatism, and other internal problems. It is amazing how a few degrees difference in temperature can alter the distribution of a plant. Today, only a few populations of bog bean survive in dune habitats that are exposed to frequent fogs and cool offshore winds. Most thriving populations of bog bean occur along the edges of mountain lakes, where the temperatures are similar to those that existed along the coast 10,000 years ago.

Aside from a snout moth caterpillar, no specific herbivores were found on some bog bean plants growing in some inland dune wetlands. It is not known if the spindly, filamentous hairs on the petals play any role in the pollination process. Bees, flies, and shining flower beetles (Coleoptera: Phalacridae) visit the flowers and are considered likely pollinators.

BOG BEAN COMMUNITY Herbivores
Bees
Flies
Shining flower beetles
Snout moth

BOG BEAN
(*Menyanthes trifoliata*: Menyanthaceae)
Characteristics: Aquatic to semi-aquatic perennial. Leaves alternate, trifoliate. Flowers whitish; five petals fringed with long curled hairs; borne on erect, leafless stalks.
Regional habitat: Select wetland habitats from Washington to California.
Comments: Bog bean is circumboreal and also occurs at high elevations in the Pacific Northwest. It survives as a relic in a few cool habitats along the coast.

SNOUT MOTH
(*Udea profundalis*: Lepidoptera: Crambidae)
Characteristics: Caterpillar pale yellow to cream; integument transparent; pronotum with two black dorsal spots. Adult with pronounced snout and mottled reddish-brown wings.
Regional habitat: From Washington to California along the Pacific Northwest coast.
Comments: This is a polyphagous moth. Bog bean is probably just one of many known host plants.

YELLOW MONKEY FLOWER

Yellow monkey flower is similar to many dune wetland plants that suddenly appear in the most unusual locations and then are never seen in that particular spot the following years. The erect and protruding flowers do resemble a monkey's head. Little red spots at the base of the petals are reminiscent of the little red mites that live in the flowers of beach morning glory, except that those spots are constantly moving. Red stems and sessile, roundish, opposite leaves are other diagnostic features of yellow monkey flower. I was surprised how rapidly the plant grew and flowered, which was fortuitous since it was rooted at the bottom of a fragile wet cliff. Even with this isolated plant, a few insects are still able to locate it.

YELLOW MONKEY FLOWER COMMUNITY **Herbivores**
Bumblebees
Sawflies
Caterpillars

YELLOW MONKEY FLOWER
(*Erythranthe guttata*: Phrymaceae)
Characteristics: Short-lived perennial with short stems, opposite roundish leaves and large two-lipped yellow flowers with red-spotted throats.
Regional habitat: Wet habitats from Washington to California.
Comments: Bumblebees visit the flowers and caterpillars develop in the seedpods of yellow monkey flower.

SAWFLY
(Hymenoptera: Cimbicidae)
Characteristics: Larva tan with irregular white line down both sides; yellow and brown spots over entire body; head brown on top, white beneath; eyes black.
Regional habitat: Larvae feed on leaves in various wet habitats along the Oregon coast.
Comments: Adults of this colorful sawfly larva were not obtained.

WATERCRESS

Watercress is certainly a well known plant of the dune wetlands, not because of its four small white petals and roundish leaves but because of its use as a source of vitamins, especially vitamin K. To some, watercress is a source of income, which is why the leaves or sprouts are commercially available. Caterpillars and beetles eat the leaves. Due to its rapid growth in a variety of semi-aquatic habitats, including slow moving creeks, watercress is sometimes considered an invasive weed and has spread far and wide from its native home in Eurasia.

WATERCRESS COMMUNITY Herbivores
Caterpillar
Banded flea beetle
Midge

WATERCRESS

(*Nasturtium officinale*: Brassicaceae)

Characteristics: Aquatic or semi-aquatic perennial. Stems creeping, rooting at nodes. Leaves numerous, smooth, divided into three to seven roundish leaflets. Flowers small, white; grouped at tips of short stalks. Seedpods elongate.

Regional habitat: Various wetlands throughout the Pacific Northwest.

Comments: Watercress is considered Eurasian in origin but is now globally distributed in temperate climates. It is cultivated for its edible greens. A caterpillar and the banded flea beetle feed on the exposed leaves. Midge larvae (Diptera: Chironomidae) develop on the underside of submerged watercress leaves.

CATERPILLAR

(Lepidoptera)

Characteristics: Caterpillar dark gray with narrow orange lateral bands running length of body; head black.

Regional habitat: On leaves of watercress in Oregon.

Comments: Adults of this caterpillar were not obtained.

BANDED FLEA BEETLE

(*Systena* sp.: Coleoptera: Chrysomelidae)

Characteristics: Larva white; develops in soil. Adult dark purple to black; elytra with paired yellow longitudinal bands.

Regional habitat: On watercress in Oregon.

Comments: This species has several different color forms. Adults feed on leaves of watercress while the larvae feed on roots of a wide range of plants.

POND LILIES

Of the two types of pond lilies found in dune wetlands, the native yellow-flowered spatterdock is the most common. In some lakes, spatterdock plants have become so abundant that the large, flat, round leaves are forced out of the water for lack of space. After admiring a cluster of spatterdock flowers in a small dune lake, I decided to wade out to search for pollinating insects.

Wearing a bathing suit and rubber shoes, I carefully started moving toward the nearest flowering plant, which was about two meters from shore. I was surprised how deeply my feet sank into the gooey mire on the bottom of the lake; it seemed like the lake bottom was composed of slush. Pulling my legs out of the goo was a challenge. Finally I made contact with a large flat spatterdock tuber. I gingerly edged along the long, slender tuber, trying to balance myself to keep from falling and becoming embedded in the sediment. The beautiful, unusual yellow flowers that I was hoping to observe seemed to be laughing at me. Finally, I got close enough to see the green sepals, yellow petals, and series of stamens in the center of the flowers. The stamens reminded me of a group of tall soldiers with red scarves and yellow caps. Honeybees hovering around the flowers appeared to be the major pollinators.

The white water lily with its sublime white petals and pristine yellow stamens looks quite aristocratic in comparison to spatterdock. There are no beetles to mar the leaves and only a few aphids and an occasional caterpillar dot the leaf surfaces. Most creatures associated with white water lily, like mites, brown limpets, and snails, are on the undersurface of the leaves and go unnoticed.

SPATTERDOCK COMMUNITY Herbivores
Honeybees
Leaf beetle
Stem midge
Snail

SPATTERDOCK
(*Nuphar lutea*: Nymphaeaceae)
Characteristics: Aquatic perennial. Tubers large, massive. Leaf stems long, arise from tubers; leaf blades round to oval, heart-shaped, floating or partly emergent. Flowers cup-shaped with green sepals and yellow petals; stamens arranged in a ring around the centrally positioned ovary.

Regional habitat: Ponds, lakes, and other dune wetlands from Washington to California.

Comments: Spatterdock (also called yellow pond lily) can become so abundant in dune lakes that the raised flat leaves are continuous over the entire water surface. Honeybees appear to be the main pollinators.

LEAF BEETLE

(*Galerucella* sp.: Coleoptera: Chrysomelidae)

Characteristics: Eggs round, light yellow, deposited in groups. Larvae black with well-defined body segments. Adults greenish-brown with yellowish pronotum bearing two or more black spots; eyes black, small, protruding from head; elytra with yellow borders.

Regional habitat: On spatterdock in Oregon and California.

Comments: Both larvae and adults feed on the leaves and, when common, can defoliate the plants.

STEM MIDGE

(Diptera: Chironomidae)

Characteristics: Maggots small, worm-like, bright red, glistening; tail with terminal segment orientated at right angles to body; head rounded. Pupa elongate, terminal segment papillate. Adult small, black.

Regional habitat: In stems of spatter-dock in Oregon.

Comments: Larvae develop inside stems and pupae are formed in the same location. The short angled terminal segment on the larvae provides better mobility within the stems of spatterdock.

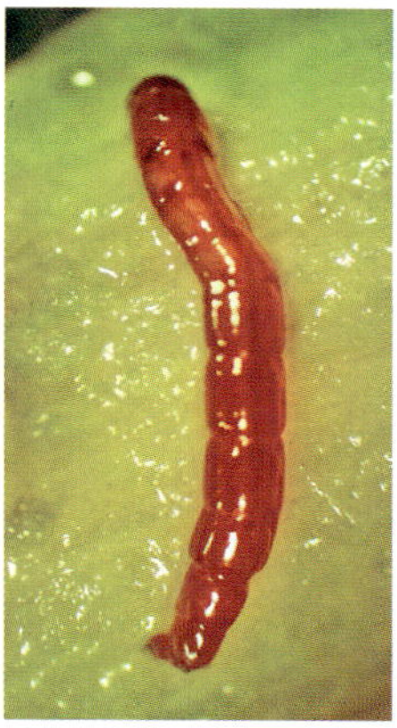

SNAIL

(Gastropoda: Limnaeidae)

Characteristics: Small, brown snail with roundish wrinkled shell; two tentacles tipped with "eyes."

Regional habitat: On spatterdock leaves in Oregon.

Comments: The snails feed on spatter-dock leaves. Stalked eyes on these snails move continuously and give the impression that the observer is being scrutinized.

WHITE WATER LILY COMMUNITY		
Herbivores	Scavengers	Parasite
Aphid	Limpet	Wasps
Leaf caterpillar	Pond snail	
	Water mite	

WHITE WATER LILY
(*Nymphaea odorata*: Nymphaeaceae)
Characteristics: Aquatic perennial. Tubers large. Leaves oval-rounded, floating or slightly submerged; petioles long, arise directly from tubers. Flowers fragrant with numerous white petals and yellow anthers.

Regional habitat: Ponds and lakes from Washington to California.
Comments: White water lily was introduced from the eastern United States and is well established along the Pacific coast.

LIMPET

(*Ferrissa* sp.: Mollusca: Planorbidae)

Characteristics: Brown, oval-shaped shell lacking whorls but covering entire body.

Regional habitat: Associated with white water lily in dune ponds and lakes in Oregon.

Comments: Limpets feed on algae and other decaying plant matter adhering to the underside of white water lily leaves. Shell edges that adhere firmly to leaf surface protect internal organs.

APHID

(*Rhopalosiphum nymphaeae*: Hemiptera: Aphididae)

Characteristics: Brownish wingless aphids with long cornicles and straight antennae held over body. Winged individuals appear in fall and spring.

Regional habitat: Associated with white water lily in dune ponds and lakes in Oregon.

Comments: This aphid has a life cycle that alternates between two different plant hosts. Summer populations are mostly wingless and occur on leaf surfaces of white water lily. Fall and spring populations are initiated when winged individuals fly from white water lilies to various fruit trees and undergo mating and oviposition. Aphids on white water lily are parasitized by small wasps.

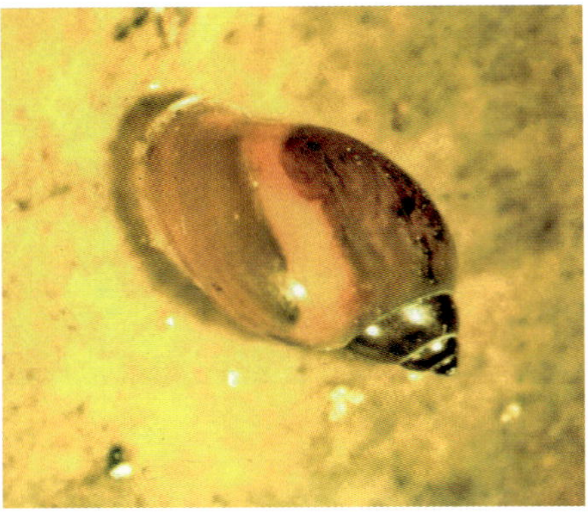

POND SNAIL

(*Physella gyrina*: Mollusca: Physidae)

Characteristics: Small snail with oval, asymmetrical, brownish shell with opening on right side when examined ventrally.

Regional habitat: Dune ponds and lakes from Washington to California.

Comments: This snail genus is indigenous and widely distributed in North America. Food consists of microorganisms adhering to the underside of white water lily leaves.

LEAF CATERPILLAR

(*Elophila* sp.: Lepidoptera: Crambidae)

Characteristics: Caterpillar dark green (matching color of white water lily leaves), with three broken reddish longitudinal lines extending length of body.

Regional habitat: Dune ponds and lakes in Oregon.

Comments: The caterpillars are voracious feeders of white water lily leaves. Pupae and adult moths were not obtained.

WATER MITE

(*Hydrozetes* sp.: Acari: Hydrozetidae)

Characteristics: Reddish-brown, tear-drop-shaped mite with pointed head, six (larva) or eight (adult) legs.

Regional habitat: Dune ponds and lakes in Oregon.

Comments: This "aquatic" mite feeds on microorganisms growing on the underside of white water lily leaves.

WATER PARSLEY

This early summer blooming perennial, which often reaches over a meter in height, is common along the borders of dune lakes and ponds from Alaska to California. It also occurs in marshy areas and other wetlands. In some cases, the base of the plant is submerged, so examining the flowers can be a wet experience. Frog tadpoles often rest under the lower leaves that dip down in the water. Medicinally, water parsley was used as a laxative and to cure upset stomachs and headaches. Whether the roots and other parts of water parsley are edible is controversial. However, since water parsley is so similar in appearance to poison hemlock, it is best not to attempt to collect and eat the roots. One curious feature of water parsley is the appearance of a second flower cluster after seeds from the first one have reddened and matured.

WATER PARSLEY COMMUNITY	
Herbivore	Predator
Sweat bee	Crab spider

WATER PARSLEY

(*Oenanthe sarmentosa*: Apiaceae)

Characteristics: Delicate perennial with weak, reclining stems. Leaves divided two or three times; leaflets toothed. Flowers white; borne in terminal clusters at stem tips. A second set of flowers may appear at tip of maturing reddish capsules of the first flower cluster.

Regional habitat: Low wet sites along streams, marshes, and lakes from Washington to California.

Comments: Various pollinators visit the flowers, some of which end up as dinner for flower crab spiders.

SWEAT BEE

(*Lasioglossum* sp.: Apidae: Halictidae)

Characteristics: Small, solitary, dull, metallic black bee with wide head and thorax with some longish white hairs; abdomen with several white transverse bands.

Regional habitat: Along the coast from Washington to northern California

Comments: Sweat bees usually nest in the ground and carry pollen in the stiff hairs on their hind legs.

CRAB SPIDER

(*Misumena vatia*: Araneae: Thomisidae)

Characteristics: Small, white to yellow spiders with a pair of orange bars on side of abdomen; first two pairs of legs elongated, opened widely at rest.

Regional habitat: Dune forests from Washington to California.

Comments: Female crab spiders rest on the flowers of water parsley and remain motionless with their front legs widely spread, ready to snap shut on a visiting insect. Sweat bees and other pollinators are the usual victims.

MARSH CINQUEFOIL

I first encountered marsh cinquefoil years ago when I was exploring the offshore Dutch island of Terschelling in the North Sea. It baffled me because, while the leaves were five-parted like most cinquefoils, the flowers were red, not the typical yellow color of other cinquefoils. Suddenly my attention was directed to a bumblebee that had settled on one of the flowers. The bee was slowly floundering around on the flower. Suddenly it fell off and began walking in circles on the ground. In this state, the bumblebee was easy to pick up and did not try to sting me. It was intoxicated by the nectar of marsh cinquefoil. A local naturalist explained to me later that nectar of this plant makes bumblebees tipsy and befuddled.

The marsh cinquefoil that occurs along the Pacific Northwest coast appears very similar to the Dutch island populations, but I have not had the opportunity to observe any bumblebees visiting it. While the red, cup shaped flowers are quite striking, they emit a fetid odor. The numerous seeds are quite small and attractive to seed bugs, which gather within the capsules to feed. It is amusing to watch the bugs moving over the dried flowers, checking to see if the seed capsule has opened enough for them to squeeze inside.

MARSH CINQUEFOIL COMMUNITY Herbivores
Thrips
Seed bug
Colletid bee

MARSH CINQUEFOIL
(*Comarum palustre*: Rosaceae)
Characteristics: Perennial with ascending stems and alternate, divided leaves. Flowers reddish-purple, bowl-shaped; borne at tips of terminal and lateral stems.

Regional habitat: Wet meadows, bogs, and lakeshores from Washington to California.

Comments: Marsh cinquefoil is circumboreal and the only cinquefoil with reddish-purple flowers. Thrips (Thysanoptera) feed on pollen in the flowers.

SEED BUG

(*Cymus* sp.: Hemiptera: Lygaeidae)

Characteristics: Small, brown, elliptical bug covered with minute punctures; eyes red, protruding; antennae short. Juveniles with short wing pads that do not cover abdomen. Adult wings long, slender, extend over abdomen.

Regional habitat: Wet meadows, bogs, and lakeshores from Washington to California.

Comments: These bugs feed on seeds inside the seed capsules of marsh cinquefoil.

COLLETID BEE

(*Colletes* sp.: Hymenoptera: Colletidae)

Characteristics: Medium-sized, short-tongued hairy bee with reddish abdomen.

Regional habitat: Dune forests and wetlands throughout the Pacific Northwest.

Comments: Colletids are ground nesting and construct larval cells lined with cellophane-like material secreted from their salivary glands. Pollen is carried within hairs located on the middle and hind legs.

BROADLEAF ARROWHEAD

Some wetland plants have starchy tubers or roots for food storage. This feature is often associated with the presence of underground stems (rhizomes). One plant with these features is broadleaf arrowhead, easily recognized by its light green arrow-shaped leaves and white flowers. Early Americans apparently collected the tubers for food. Arrowhead actually can have three different leaf types depending on its habitat. Plants growing in deep water may also bear submersed and floating leaves along with the more visible aerial leaves.

BROADLEAF ARROWHEAD
(*Sagittaria latifolia*: Alismataceae)
Characteristics: Semiaquatic tuberous perennial. Basal leaves broad; leaf blades long-stalked, arrow-shaped. Flowers with three white petals; borne in groups of three on leafless stems. Male flowers with center cluster of yellow stamens at top of flower stem. Female flowers with developing seeds in bottom leaf whorl.
Regional habitat: Marshes, ponds, and lakes from Washington to California.
Comments: The tubers are very starchy and are eaten by waterfowl.

HALOPHYTE PLANTS

While halophytic, or salt-tolerant, plants are common in tidal flats and along the edges of salt marshes and ponds near the sea, they can also occur inland in high saline habitats. Growing in saline environments provides little competition from other vegetation and the high concentration of salt in their tissues makes them unpalatable to most animals. In the Pacific Northwest, three such plants are fleshy jaumea, pickleweed, and seablite. The succulent leaves and stems of these halophytes are cooked and eaten. A tea made from boiled leaves of fleshy jaumea is said to cure fevers. Perhaps the high salt content is why there are not many insects feeding on these halophytes. Only meadow katydids could be found feeding on pickleweed and fleshy jaumea.

FLESHY JAUMEA

(*Jaumea carnosa*: Asteraceae)

Characteristics: Creeping perennial with opposite, succulent, narrow leaves. Flowers small, yellow; borne singly at tips of short stalks arising from stems.

Regional habitat: Coastal salt marshes and other wet, saline areas from Washington to California.

Comments: Fleshy jaumea, which also occurs along the western coast of South America, contains arbuscular mycorrhiza in its roots. The only insect found on fleshy jaumea was the meadow katydid (*Conocephalus* sp.: Orthoptera: Tettigoniidae) (see photo under Pickleweed).

PICKLEWEED Herbivore	Parasite
Meadow katydid	Dodder

PICKLEWEED

(*Salicornia virginica*: Amaranthaceae)

Characteristics: Matted perennial. Stems numerous, prostrate to ascending and erect, jointed, fleshy. Flowers minute, in groups of three; borne in short, compact terminal spikes along flower stalk tips.

Regional habitat: Marshes and tide flats from Washington to California.

Comments: Meadow katydids were the only insects found feeding on pickleweed. The stems, which are occasionally made into salads, are heavily parasitized by dodder.

DODDER

(*Cuscuta* sp.: Convolvulaceae)

Characteristics: Parasitic flowering plant lacking chlorophyll. Stem yellowish, thin, wiry, winds around host; small protuberances adhere to and penetrate host stems. Leaves minute, scale-like, barely visible. Flowers relatively large with five white, pointed petals.

Regional habitat: On various plants throughout the Pacific Northwest.

Comments: Dodder was presented previously as a parasite of water hemlock. It is only mentioned again because of its abundance on pickleweed. The young plants sustain themselves on nutrients carried over from the seeds until they make contact with pickleweed. At that point, dodder becomes completely parasitic and survives from nutrients taken from its host.

MEADOW KATYDID

(*Conocephalus* sp.: Orthoptera: Tettigoniidae)

Characteristic: Large, greenish insect with tan eyes and forewings that are as long as or slightly longer than the abdomen. Female (shown here) has long sword-like ovipositor protruding from tip of abdomen.

Regional habitat: On pickleweed and other marsh plants throughout the Pacific Northwest.

Comments: The coloration of meadow katydids makes it difficult to find them among the numerous stems of pickle-weed. The meadow katydid also feeds on fleshy jaumea.

SEABLITE

(*Suaeda maritima:* Amaranthaceae)

Characteristics: Annual halophyte with numerous, branched, fleshy, ascending to erect stems. Leaves pointed, succulent. Flowers small, whitish, clustered in leaf axils. Fruit reddish, berry-like.

Regional habitat: Marshes, tide flats, and along the borders of ponds near the sea from Washington to California.

Comments: The leaves can become red with an excess of salt. Curiously, when dried, the salt turns the leaves black, thus the origin of the Latin name "Suaeda" that means "black salt" in Arabic.

GOLDEN-EYED GRASS

This small perennial, which is usually less than a foot in height, bears clusters of yellow flowers that appear in May. It prefers wet areas and often occurs in marshes, wet meadows, and along the edges of dune ponds and lakes. The bright orange anthers, which attract bees, are a contrasting color to the yellow sepals and petals. The seeds are borne in egg-shaped capsules that turn dark brown at maturity.

GOLDEN-EYED GRASS
(*Sisyrinchium californicum*: Iridaceae)
Characteristics: Perennial. Leaves narrow, arising from base of plant. Flowers borne singly or in small clusters at tips of winged stems. The yellow sepals and petals, which are identical in size and shape, have five dark veins on their surfaces.

Regional habitat: Moist areas, bogs, ponds, and along seepage edges from Washington to California.
Comments: The yellow flowers are about the same color as those of the yellow monkey flower.

BABY BLUE EYES

It is surprising to see divided leaves on this little annual with flowers that dwarf the rest of the plant. The flower color is variable, ranging from white to blue or light purple in the wild but some commercial varieties have variegated petals. The five petals often contain faint veins that are dark spotted, especially near the bases. There are several varieties of baby blue eyes that occur in different habitats ranging from the coast to mountaintops. Its ability to tolerate such a range of habitats and climate changes makes it an ideal plant for the home garden.

BABY BLUE EYES
(*Nemophila menziesii*: Boraginaceae)
Characteristics: Small, fragile plant with basal leaves pinnately divided into short lobes. Five pinkish-blue petals with darker spotted veins and roundish tips.

Regional habitat: Wet meadows along edges of dune forests in Oregon and California.
Comments: There is a fair amount of color variation of the petals, even between flowers borne on different flower stalks of the same plant.

NARROWLEAF BUR-REED

Narrowleaf bur-reed is another perennial that emerges each year from rhizomes. The long, narrow leaves may be submersed in water, floating, or emerged and surrounding the flower stalks that can reach over a meter in height. These strange flowers have no noticeable sepals or petals. Clusters of straight white stamens on the male flowers resemble conidiophores of some fungi. Female flowers are positioned beneath to catch pollen falling from male flowers.

NARROWLEAF BUR-REED
(*Sparganium angustifolium*: Typhaceae)
Characteristics: Perennial. Leaves long, slender, basal. Flowers whitish; male and female flowers borne in separate spherical heads on same plant; male flowers possess elongate stamens while female flowers, positioned below the male flowers, have short stigmas, as shown.
Regional habitat: Lakes and ponds from Washington to California.

Comments: The plants can grow partially or completely submerged. Insects feed upon the seed clusters during the night **(B)**. The floating plant in the water behind narrowleaf bur-reed **(A)** is an aquatic fern of the genus *Marsilea* (Marsileaceae). This non-native genus of ferns has been introduced into some Pacific Northwest aquatic habitats.

SEA ARROW GRASS

Brackish marshes along the edges of dune forests are the habitat for sea arrow grass, a perennial with a stout, starchy rhizome. The flower stalks are straight and slender, reaching up to a meter in height. Tiny flowers closely attached to the stalks are not impressive, yet this humble little plant was part of the late glacial vegetation that provided food for woolly mammoths. While those behemoths are now extinct, sea arrow grass still occurs throughout the Pacific Northwest. One survival feature is having seeds engineered to attach to feathers and be spread by waterfowl. Since sea arrow grass is said to be poisonous to livestock, I wonder what affect it had on grazing mammoths.

SEA ARROW GRASS
(*Triglochin maritima*: Juncaginaceae)
Characteristics: Perennial. Leaves long, slender, basal. Flower stalks green, leafless. Flowers inconspicuous; borne in a series along stalk. Mature seed capsules and stalk light brown.
Regional habitat: Wetlands in the Pacific Northwest from Washington to California.
Comments: The seeds are carried by wildfowl, which may explain why sea arrow grass is circumboreal.

LIZARD'S TAIL

It was a surprise to find this native eastern North American perennial in the Pacific Northwest, but the extended flower spike with the recurved tip and heart-shaped leaves readily identifies lizard's tail. Appropriately, the common name is based on the curved flower stalk that resembles a lizard's tail. This plant favors wet habitats and occurs around the edges of dune lakes. The rhizomatous nature of lizard's tail allows it to form dense clumps that ducks and other wildlife use for concealment. While the foliage is considered toxic to mammalian herbivores, tea from dried leaves has been used to treat inflamed areas of the skin.

LIZARD'S TAIL
(*Saururus cernuus*: Saururaceae)
Characteristics: Tall perennial. Stalks upright to tapering. Leaves, large, triangular. Flowers small, white; in tightly placed series along stalk.

Regional habitat: Wetlands along the edges of dune forests in Oregon and California.
Comments: This plant is native to eastern North America but has become naturalized along the Pacific coast.

PICKEREL WEED

The erect spike of blue flowers of pickerel weed is often adjacent to a single, wide leaf that reflects sunlight, thus serving as a living backlight. This herbaceous emergent aquatic perennial loves the sun and usually grows in dense colonies with the shiny arrowhead-shaped leaves rising above the water surface. The color of the flowers is somewhat variable, with occasional white and purple blooms appearing. Fish collect under the protection of the leaves, thus the name pickerel weed.

Deer also feed on the leaves that are close to shore and wildlife as well as humans consume the seeds. Young leaves have been used in salads.

PICKEREL WEED
(*Pontederia cordata*: Pontederiaceae)
Characteristics: Tall emergent aquatic. Leaves heart-shaped; one is positioned behind each flower stalk. Flowers blue; blooming in a series on the stalk from bottom upwards.
Regional habitat: Shallow, quiet waters along the edges of dune forests in Oregon and California.
Comments: This is another eastern North American native that has become naturalized in the West.

FLOATING PONDWEED

Floating pondweed, which occurs throughout the world, is one of the most widespread perennial plants of dune wetlands. Usually only the broadly ovate floating leaves are visible, some of which protrude with their long stalks above the water surface. Floating pondweed spreads vegetatively by runners but also forms egg-shaped seeds that contain air spaces that keep them buoyant. Floating pondweed is dioecious with male and female sexes on separate plants. The flowers are wind pollinated. While floating pondweed can provide food and a sanctuary for wildlife, it can become invasive by blocking irrigation channels, affecting boat trafficking, and hindering fishing operations. The roots and stems are said to be edible and the leaves have been used to reduce fever.

FLOATING PONDWEED
(*Potamogeton natans*:
Potamogetonaceae)
Characteristics: Two leaf types: 1) submersed, linear (actually modified leaf stalks); 2) floating, large, long-stalked, broad, lustrous, coppery-green. Flowers tiny; borne on long straight spikes that emerge from water surface.

Regional habitat: Shallow waters along the edges of dune forests from Washington to California.
Comments: This species is cosmopolitan in the Northern Hemisphere. The various metallic hues of the rusty glossy leaves look like they were painted.

CATTAIL

Dark brown seed heads at the tips of tall stems in wetlands are the sign of cattails, one of the most common marsh plants throughout North America. The appearance of green, strap-like leaves rising from the rotting vegetation of the previous year is an early announcement of spring. Soon after the new flower stalks emerge between the leaves, the red-winged blackbirds arrive. The male blackbirds, with their red and yellow shoulder patches, arrive first and preen themselves as they wait for the females to appear. The timid brown and white mottled females are well camouflaged and don't look like they are related to the males. Dried cattail leaves are used for nest construction. Hatching of the chicks is timed with flowering of the cattails. This allows the females to strip away developing seeds and pick up associated insects. The young birds are uniformly gray at first, resembling neither parent, and only later acquire their adult colors.

It is difficult to find another plant that provides so many uses to so many animals, including humans. Cattail canes, leaves, flower stalks, underground stems (rhizomes), and pollen are eaten. Humans made paper and fiber from stems, ethanol from starch in the tubers, small boats from stems, and life vests and bedding from flower down.

Removing the male portion of the flower stalk exposes numerous small seed bugs feeding on the pollen. When the stamens dry up, the bugs move down into the female portion of the flower spike and feed on developing seeds. Here they encounter green cattail moth caterpillars and picture wing fly maggots.

BROAD-LEAVED CATTAIL COMMUNITY **Herbivores**
Cattail moth
Picture wing fly
Seed bug
Red-winged blackbird

BROAD-LEAVED CATTAIL

(*Typha latifolia*: Typhaceae)

Characteristics: Tall erect perennial. Leaves flat, narrow. Flowers numerous; borne in terminal spikes; yellowish male flowers positioned above brownish female flowers. Male flowers soon disintegrate, leaving maturing female flowers.

Regional habitat: Marshes, ponds, and lakes from Washington to California.

Comments: Both rhizomes and pollen of cattails served as human food and the seed fluff was used to stuff pillows.

CATTAIL MOTH

(*Argyrotaenia* sp.: Lepidoptera: Tortricidae)

Characteristics: Caterpillar light green with narrow white rings widely spaced over body; head light brown. Adult tan; wings with fine, dark brown, irregular transverse lines.

Regional habitat: Wetlands of Oregon and California.

Comments: Caterpillars and pupae are deeply buried in female flower heads.

PICTURE WING FLY

(*Chaetopsis massyla*: Diptera: Ulidiidae)

Characteristics: Adult small, dark green, metallic; eyes large, tan to reddish; wings with three dark bands, the two apical bands joined at wing margin.

Regional habitat: Marshlands of Oregon and California

Comments: The maggots feed on decaying cattail tissue previously infested by cattail moth caterpillars.

SEED BUG

(*Kleidocerys* sp.: Hemiptera: Lygaeidae)

Characteristics: Reddish-brown with fine pits and protuberances; head triangular; eyes protruding; antennae short.

Regional habitat: Wetlands from Washington to California.

Comments: These bugs bury themselves in mature flower heads of broad-leaved cattail and feed first on pollen and then on developing seeds.

RED-WINGED BLACKBIRD

(*Agelaius phoeniceus*: Icteridae)

Characteristics: Female brown; wing lines whitish; short dark smudge behind eye. Male black; red and yellow shoulder patches can be exposed or concealed voluntarily.

Regional habitat: Breeding populations occur throughout the Pacific Northwest.

Comments: Red-winged blackbirds are omnivorous, feeding on cattail seeds and associated insects. The birds often spend their entire lives around cattails, raising their young in basket-shaped nests attached to cattail stalks. Birds infected with Lyme disease pathogens have been found in California.

AMERICAN BROOKLIME

If you find a small fragile aquatic perennial with opposite, smooth-toothed leaves and flowers with four or five blue petals and only two stamens, it certainly is American brooklime. Frequently growing in water with a depth almost reaching the level of the flowers, it often grows in dense clusters. The extended stems and rhizomes are frequently carried downstream during storms, which is why American brooklime is so widespread.

AMERICAN BROOKLIME
(*Veronica americana*: Plantaginaceae)
Characteristics: Small semi-aquatic perennial forming creeping tubers or trailing stems. Leaves narrow, toothed, opposite. Flowers small, delicate, blue to violet; borne in terminal or axillary racemes or spikes.

Regional habitat: Wetlands, especially along streams, from Washington to California.
Comments: The leaves of American brooklime are used in salads. The plant is said to have medicinal properties, although the root may contain toxins.

PARROT FEATHER

The delicate lacy leaves of parrot feather can be seen in many dune ponds and lakes. This is the only plant in the Pacific Northwest dune wetlands that originates from the Amazon River in Brazil. In spite of its use in sheltering wildlife, parrot feather, which is named after its feather-like leaves, crowds out native flora and is considered a pest in many areas. While flea beetles feed on parrot feather in Florida, no insects were found on the plants I examined in the Pacific Northwest dune wetlands.

PARROT FEATHER
(*Myriophyllum aquaticum*: Haloragaceae)
Characteristics: Aquatic perennial. Leaves whorled, feather-like. Stems long, spreading. Flowers minute, white to pinkish-white, commonly found in emergent leaf axils.

Regional habitat: Widespread in many aquatic habitats along the Pacific Northwest.

Comments: Parrot feather was introduced from South America and has been quite invasive, although the fruits and leaves are eaten by waterfowl.

DUNE WETLAND VERTEBRATES

A number of vertebrates occur in Pacific Northwest wetlands and some of the more obvious ones that naturalists are likely to see as they explore plant communities are presented below.

AMERICAN BEAVER

(*Castor canadensis*: Castoridae)

Characteristics: Large, brownish mammal with small ears; black, flattened, naked tail and webbed hind feet. Front feet with claws, front incisors large.

Regional habitat: Lakes and ponds throughout the Pacific Northwest.

Comments: American beaver is the largest rodent in North America. Trees felled by beavers show marks of their incisors. Nests for rearing young are constructed in burrows and lodges. A close associate of American beaver is the beaver beetle.

BEAVER BEETLE

(*Platypsyllus castoris*: Coleoptera: Leiodidae)

Characteristics: Very small, brown, shiny, flattened, oval, blind beetle with rounded head; antennae two-segmented, held against head; elytra short; flight wings absent.

Regional habitat: Throughout the the Pacific Northwest.

Comments: Both larvae and adult beaver beetles feed on exudates and cells from beaver skin. Pupae occur in soil at the top of the beaver dam or lodge.

BARN SWALLOW

(*Hirundo rustica*: Hirundinidae)

Characteristics: Deeply forked tail is key identifier. Blue to black above, cinnamon-tan below; bill short; feet tiny; wings pointed.

Regional habitat: Widespread and often seen catching insects on the wing over dune aquatic habitats, especially ponds and lakes, from Washington to California.

Comments: Barn swallows are nearly cosmopolitan and often rest on wires. The nest is often constructed in barns and human dwellings. They are attacked by parasitic mites (Acarina) and bugs (Hemiptera).

WESTERN GULL
(*Larus occidentalis*: Laridae)
Characteristics: Large bird with dark back; under parts white; bill yellow with spot of red on undersurface near tip; feet pinkish.
Regional habitat: Various dune aquatic habitats, especially ponds and lakes, from Washington to California.
Comments: Western gulls sometimes frequent water bodies far from the sea, like this one in a dune lake.

CANADA GOOSE

(*Branta canadensis*: Anatidae)

Characteristics: Large bird with black head, white chin strap and white sides with the latter most conspicuous when resting on water.

Regional habitat: Lakes and ponds in the Pacific Northwest.

Comments: Groups of Canadian geese settle on dune lakes in the spring, but usually continue north to search for nesting sites.

BELTED KINGFISHER

(*Megaceryle alcyon*: Alcedinidae)

Characteristics: Dark blue bird with a blue breast band, crested head and long, straight bill. Females with additional ragged, rusty, breast band (as in photo).

Regional habitat: Along water sources throughout Pacific Northwest.

Comments: This bird hovers above the water until it sees a fish, then plunges straight down and catches it in its large bill. It likes to perch on telephone wires to survey the surroundings.

GREAT BLUE HERON

(*Ardea herodias*: Ardeidae)

Characteristics: Large bluish wading bird with long, folded neck; head black on top with white lateral line passing over eyes; bill large, sharply-pointed; legs long, thin.

Regional habitat: Marshes, lakes, as well as saline bays throughout Pacific Northwest.

Comments: Hunting food is accomplished by wading slowly through the water, then suddenly plunging the head down to catch a small fish or crustacean.

GREAT EGRET

(*Ardea alba*: Ardeidae)

Characteristics: Tall, slender, white bird with a long neck, narrow, yellow, pointed bill and black legs and feet.

Regional habitat: Ponds and marshes in Oregon and California.

Comments: Food and habits of the great egret are very similar to those of the great blue heron. They can often be found feeding together in dune lakes.

HOODED MERGANSER

(*Lophodytes cucullatus*: Anatidae)

Characteristics: Male head black with white crest. Female head brownish with tan crest. Crest can be raised and lowered. Juveniles, such as this immature female, lack a well-developed crest.

Regional habitat: Forest lakes and ponds in the Pacific Northwest.

Comments: Hooded mergansers are fish eaters and often dive in the water to pursue their prey.

MALLARD

(*Anas platyrhynchos*: Anatidae)

Characteristics: Male mostly white with metallic green head and neck, separated from brown breast by white ring; bill yellow. Female mottled brown with orange bill.

Regional habitat: In forest ponds and lakes throughout the Pacific Northwest.

Comments: Mallards will often visit small, confined water sources avoided by other waterfowl.

BROOK TROUT
(*Salvelinus fontinalis*: Salmonidae)
Characteristics: Greenish-gray body covered with numerous small white dots; sides with series of faint darker vertical blotches; dorsal fin with rows of broken dark lines.
Regional habitat: Ponds and streams in the Pacific Northwest.
Comments: Brook trout is native to eastern North America and has been introduced in Pacific Northwest ponds and lakes as a food and game fish. The only native trout in the Pacific Northwest is a race of rainbow trout that spends its life in fresh water. Another race of rainbow trout that lives in the ocean and only enters freshwater streams to spawn is the steelhead trout. Other game fish introduced from eastern North America and occasionally encountered in dune lakes and ponds are largemouth bass, common shiner, sunfish, and yellow perch.

PACIFIC TREE FROG

(*Pseudacris regilla*: Hylidae)

Characteristics: Small green and brown frog, often with areas of reddish-brown on back; head with black eye stripe; toes equipped with pads. Tadpoles black, gills internal.

Regional habitat: Various habitats throughout the Pacific Northwest.

Comments: On wet days, Pacific tree frogs can be found resting on plant leaves. In the photo, the frog is sitting on a thimbleberry (*Rubus parviflorus*) leaf. Tadpoles can develop in a variety of aquatic habitats, including some with a high salt content. The hind legs form before the front legs. The long tail is gradually diminished in size, then lost and transformation to the adult is complete. These frogs can produce a loud, high-pitched, penetrating screech.

ROUGH-SKINNED NEWT

(*Taricha granulosa*: Salamandridae)

Characteristics: Brown above, lighter below; skin smooth (not rough and pitted as in terrestrial phase); eyes small; tail laterally flattened; gills external, well-visible at back of head.

Regional habitat: Forested and wetland habitats from Washington to California.

Comments: This is the aquatic phase of the rough-skinned newt. Larvae occur in various dune wetland habitats, especially ponds, marshes, and small lakes. They feed on earthworms and developing stages of aquatic invertebrates. As larvae mature, the gills shrink and finally fall off, the lungs taking over the role of respiration. Both adults and larvae of rough-skinned newts are poisonous and should be handled with care.

DUNE WETLAND INVERTEBRATES

ALDERFLY

(*Sialis* sp.: Neuroptera: Sialidae)

Characteristics: Larva aquatic with extended abdomen bearing a series of protruding fringed tracheal gills. Adult slender, brown to tan; antennae long, narrow; wings long, covering abdomen and held roof-like over body.

Regional habitat: Various aquatic habitats throughout Pacific Northwest.

Comments: Adults are active during daylight, and when pursued, will often run rather than fly. Larvae are predaceous and conceal themselves in bottom ooze waiting for prey.

BLACK STONE FLY

(*Perla* sp.: Plecoptera: Perlidae)

Characteristics: Nymph aquatic, dark brown with two tail filaments, wide head, large mandibles, short antennae and reduced gills located at leg bases. Adult black; eyes protrude from sides of head; tail filaments short, not extending beyond wings; antennae long; wings held flat over abdomen.

Regional habitat: Aquatic habitats, especially in flowing water, throughout Pacific Northwest.

Comments: Adults take no or little nourishment and often rest on stones adjacent to water sources. Nymphs are omnivores but prefer to eat plant detritus. Their bodies are often flattened and closely attached to the undersurface of stones.

PREDACEOUS DIVING BEETLE

(*Cybister* sp.: Coleoptera: Dytiscidae)

Characteristics: Larva brown to black; body attenuated; terminal abdominal segment with paired tail filaments; head large; mandibles sickle-shaped. Adult large, oval, shiny, elytra black or green.

Regional habitat: Aquatic habitats throughout Pacific Northwest.

Comments: Both larvae and adults are aquatic. Larvae are called water tigers due to their attacks on other aquatic insects. Hollow mandibles allow a diving beetle larva to firmly hold its victim while sucking out its body contents.

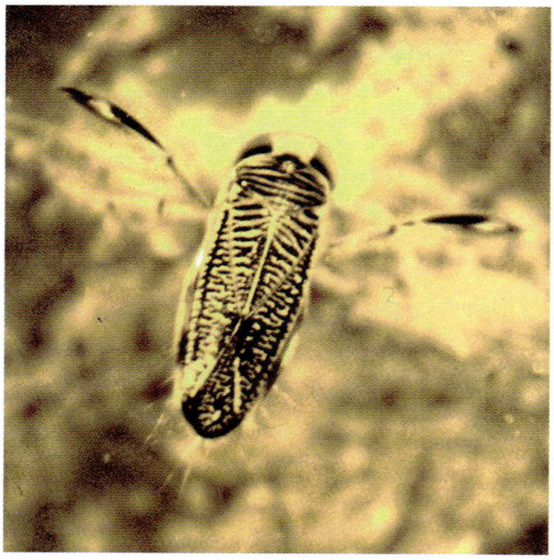

WATER BOATMAN

(*Sigara washingtonensis*: Hemiptera: Corixidae)

Characteristics: Small, oval-shaped; dorsum covered with irregular transverse bands; head rounded; eyes large; hind legs long, used as oars for swimming.

Regional habitat: Variety of aquatic habitats in Pacific Northwest.

Comments: Water boatmen are omnivores and mostly bottom feeders. Many species can tolerate saline environments. They lack gills and carry a covering of air when they descend, grasping a plant stem to keep from surfacing.

CARDINAL MEADOWHAWK
(*Sympetrum illotum*: Odonata: Libellulidae)
Characteristics: Body and wing base red (more distinct in male); thorax with two white lateral patches on each side; eyes large.
Regional habitat: Naiads (larvae) occur in lakes and ponds throughout the Pacific Northwest.
Comments: This is a widespread dragonfly that is able to develop in quite small ponds. Adults roam far and are often found some distance from their breeding sites.

VARIEGATED MEADOWHAWK
(*Sympetrum corruptum*: Odonata: Libellulidae)
Characteristics: Head and thorax gray; thorax with two oblique white lines on each side ending in small yellow circle; abdomen orange with white central longitudinal stripe and rows of irregular white bands on both sides; wings clear, outer marginal veins orange-yellow; pterostigma orange to tan.
Regional habitat: Wetlands throughout the Pacific Northwest.
Comments: Variegated meadowhawks like to perch on the tips of twigs. They are one of the few dragonflies that undergo mass migrations along the Pacific coast. Migrating populations initiated from Canada and Washington fly south in late summer, often reaching Baja California.

PACIFIC BLUE DAMSELFLY

(*Ischnura* sp.: Odonata: Coenagrionidae)

Characteristics: Mostly blue damselfly with a black band connecting the eyes; thorax with longitudinal black bars; abdomen with wide black rings alternating with small blue rings; tail tip black; wings held together over back at rest.

Regional habitat: Around ponds and lakes throughout the Pacific Northwest.

Comments: These beautiful damselflies can be easily approached when they are sunning themselves near a water source.

WATER STRIDER

(*Gerris* sp.: Hemiptera: Gerridae)

Characteristics: Slender bug with protruding eyes, sharp beak and short antennae; legs long, thin; front legs shorter than others; wings present or absent.

Regional habitat: In many different water sources in the Pacific Northwest.

Comments: These insects stride over the water surface without breaking the water film. They are predators and feed on small arthropods that fall in the water.

WHIRLIGIG BEETLE

(*Gyrinus* sp.: Coleoptera: Gyrinidae)

Characteristics: Aquatic, with elongate larva possessing numerous, fringed gills. Adult black, shiny, oval; eyes divided (upper half above water, lower half submerged); elytra with rows of minute pits.

Regional habitat: In many different water sources in the Pacific Northwest.

Comments: Both larvae and adults are predators. Adults swim rapidly in circles and are difficult to capture.

WESTERN ENCEPHALITIS MOSQUITO
(*Culex tarsalis*: Diptera: Culicidae)
Characteristics: Larva (wiggler) and pupa aquatic. Adult with small head bearing proboscis; legs long, slender; tarsi with white bars.
Regional habitat: Widespread in the Pacific Northwest.
Comments: This species feeds on a variety of birds and mammals. It vectors several human virus pathogens including western equine encephalitis, St. Louis encephalitis, and West Nile. The adults also visit flowers for nectar.

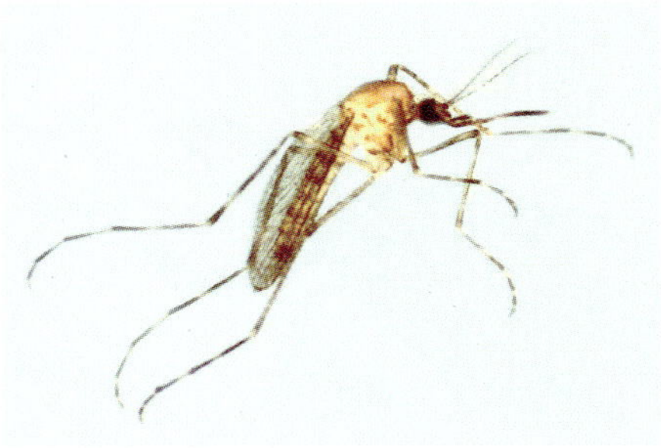

RUSTY CADDIS FLY
(*Psychoglypha* sp.: Trichoptera: Limnephilidae)
Characteristics: Adult large, reddish-brown, moth-shaped (but covered with hairs, not scales); head small; antennae long, outstretched. Larva slender, lives inside a case it constructs from bits of debris that fell in the water.
Regional habitat: In running and still aquatic habitats in the Pacific Northwest.
Comments: These cold water caddis flies are commonly used as bait by fishermen. Cases constructed by the larvae for protection can be quite bizarre and mimic the habitat of the surroundings. This one shows that spruce trees overshadowed the water source.

LONGHORN CADDIS FLY

(*Mystacides* sp.: Trichoptera: Leptoceridae)

Characteristics: Larva aquatic; abdomen enclosed in case made from variety of materials. Adult greenish to almost black; first four maxillary palps quite hairy, terminal palp bare; antennae very long.

Regional habitat: Ponds and lakes along the Pacific Northwest.

Comments: The large bushy maxillary palps identify members of this genus.

WATER SPRINGTAIL

(*Sminthurides* sp.: Collembola: Sminthurididae)

Characteristics: Tiny, globular hexapod with forked tail piece (fulcrum) at end of abdomen; antennae as long as legs; wingless.

Regional habitat: Edges of calm water bodies throughout Pacific Northwest.

Comments: These terrestrial hexapods often congregate on water surfaces. Their light bodies do not break the water film. The fulcrum, when forced downwards, propels the animal into the air.

GLOSSY-WING MAYFLY

(*Paraleptophlebia* sp.: Ephemeroptera: Leptophlebiidae)

Characteristics: Larva (naiad) aquatic, herbivorous, gilled. Adult small, delicate, slender, with long, thin tail filaments and large glossy wings with many veins.

Regional habitat: Widespread in wet habitats throughout the Pacific Northwest.

Comments: Adults commonly occur in swarms over water sources. Since the adults do not feed, their lifespan is quite short and measured in hours while that of the larva lasts for weeks or months.

MONONCHID NEMATODE

(Nematoda: Mononchidae)

Characteristics: Small, worm-like, smooth animals lacking segments and appendages; huge mouth cavity (stoma) leads to elongate, muscular pharynx.

Regional habitat: Dune water sources from Washington to California.

Comments: Mononchid nematodes are predators, catching and crushing small invertebrates in their large mouth cavities. Many other types of aquatic nematodes occur in dune water sources.

LAND PLANARIA

(Tricladida: Geoplanidae)

Characteristics: Small, reddish-brown, smooth worm-like creatures lacking segments and appendages; mouth hidden from view.

Regional habitat: Wetlands from Washington to California.

Comments: Members of this family are found in moist habitats and are predators of other invertebrates. Planarians are curious animals that have the ability to regenerate severed body portions.

GLOSSARY

Abdomen—third or terminal body portion of an insect.

Achene—a dry seed.

Alien—a species introduced from another country.

Annual—growing from seed to maturity and death in one season.

Antennae—paired sensory organs or "feelers" attached to the head of an insect.

Anther—pollen bearing part of the stamen.

Apical—outermost tip.

Arbuscular mycorrhiza—a fungus that penetrates root cells of a higher plant in a beneficial association.

Associate—an organism that is only loosely connected with another organism.

Astringent—a substance that causes contractions of tissues, especially the skin.

Back dunes—area just inland from the sand dunes, before the dune forests begin.

Bark beetle—a small beetle living and feeding within the bark or wood of plants.

Basidium—microscopic fungal structure bearing a basidiospore.

Basidiospore—spore of a basidiomycete fungus.

Beach—area between high and low tides.

Biennial—growing from seed to maturity and death in two seasons.

Biological control—a practice using parasites, predators, and pathogens to control noxious organisms.

Bog—wet soft ground that cannot support a large weight.

Broadleaf tree—a hardwood tree with wide leaves.

Carnivorous—predatory, consuming other creatures.

Carrion—decaying animal flesh.

Caterpillar—active juvenile stage between egg and pupa of a moth or butterfly.

Catkin—a scaly-bracted row of flowers, as in willows.

Ciliate—Microscopic oval protozoa covered with short cilia (hairs).

Coast—land along the edge of the sea.

Coastal dunes—accumulation of sand extending inland above the high tide zone.

Cocoon—a protective case, often hairy or silky, spun by the larva just before pupation.

Commensal—a member of an association with two organisms where one benefits and the other is not harmed.

Community—a group of interdependent organisms of different species living together in a specified habitat.

Conifer—tree characterized by needle or scale-shaped leaves and pollen and seeds borne in cones.

Cornicles—tubular structures on the abdomen of aphids.

Deciduous—a tree that sheds it leaves annually.

Deflation plain—a relatively flat area separated from the ocean by one or several rows of large sand dunes.

Disease—a disturbance in the function or structure of an organ or tissue over time.

Detritivore—an organism that feeds on dead plant and/or animal matter.

Diurnal—active in the daytime.

Dorsal or **dorsum**—the back or top.

Dune—sandy area adjacent to the sea or in deserts and other sandy localities.

Dune forest—a moist, cool climate forest growing on soil with a sandy base not far from the sea.

Dune lake—freshwater lake occurring in or adjacent to dune forests.

Dune wetlands—various fresh or slightly saline water sources (ponds, lakes, bogs, seeps, marshes, and swamps) formed on sandy soil near the sea. Often located within or adjacent to dune forests.

Ectoparasite—a parasite that lives on, rather than inside, the host.

Entire—margin continuous and smooth.

Elytra—modified first pair of wings on beetles, usually hardened to protect the underlying membranous second pair of wings and the abdomen.

Evergreen—a tree that retains living, green leaves throughout the year, eventually losing them singly.

Flower—an axis bearing one or more stamens (staminate flower), one or more pistils (pistillate flower), or both (perfect flower).

Fruit—ripened ovary with associated floral parts.

Fungus (pl. fungi)—organisms that lack chlorophyll and live as saprophytes or parasites.

Gall—abnormal plant growth caused by insect or mite activity.

Genus—a group of organisms (species) that share common morphological and genetic characters.

Global—worldwide.

Gymnosperm—plant with seeds exposed in various types of cones.

Halophyte—plant growing in saline habitat.

Hardwood—a broadleaf tree member of the angiosperms.

Haustoria—small suckers of a parasitic plant that penetrate the tissues of a host plant to withdraw water and nutrients.

Herb—plant lacking woody firm structure.

Herbaceous—seed-bearing plants that lack woody stems and die back every year.

Herbivores—organisms feeding on living plants.

Homotypic agglutination—aggregation of individuals of a species.

Host—an organism that is used as food by a parasite or pathogen.

Hypha (pl. hyphae)—microscopic cellular strands of fungi.

Infection—the process of a parasite or pathogen invading a host.

Inquiline—an animal (insect) living in the same niche or habitat as another animal. Neither benefits or suffers from the association.

Instar—stage of an insect between molts.

Integument—outer body covering of an arthropod.

Invasive species—an introduced animal or plant that adapts to a new environment and out-competes native species.

Lake—a large body of still water.

Larva (pl. larvae)—immature feeding stage between egg and pupa; e.g., caterpillar in the case of moths and butterflies and maggot regarding flies.

Legume—simple fruit dehiscing on both sutures (edges), like that of a pea.

Lichen—small, primitive plant that grows on the surface of rocks, shrubs, and trees. Each plant contains an alga and a fungus in symbiosis.

Life cycle—the various developmental stages in the life of an organism.

Litter—decomposing leaves and wood on the forest floor.

Maggot—legless, soft, juvenile stage of a fly.

Mandibles—lateral biting structures of mouthparts.

Marsh—permanent wet area where water is slightly above ground level.

Midge—small fly with long segmented antennae.

Mimic—a plant or animal that imitates another to escape predators.

Miner—larval insect developing inside plant tissue.

Mutualism—an association between two organisms where both benefit and neither is harmed.

Mycelium—strands of hyphae, the vegetative structure of a fungus.

Naiad—aquatic developmental stage of certain insects.

Native—an animal or plant that occurs naturally in a region.

Nectar—sugary deposit produced by flowers to attract insect pollinators.

Nectary—a nectar-secreting gland.

Nematode—a microscopic roundworm lacking body segments and appendages.

Nymph—immature stage of an insect that does not change in shape as it grows.

Ocelli—simple eyes, often in groups of three, on an insect's head.

Omnivore—an organism that obtains nourishment from both animals and plants.

Ovary—the ovule or seed-producing part of the flower, usually positioned in or near the middle of the flower.

Ovipositor—egg-laying apparatus in insects. It can be modified to serve as a "stinger" in some insects such as bees and yellow jackets.

Palpi—pair of segmented appendages surrounding the mouth of an insect.

Parasite—an organism that feeds in or on another organism (the host).

Parasitism—an association between two organisms where one is harmed (usually by the removal of body tissues) and the other benefits.

Parotid gland—a large salivary gland located in front of the ear.

Pathogen—a microorganism capable of causing disease in another organism.

Peduncle—stem of a flower cluster.

Perennial—a plant that survives from three to many years.

Perithecia—a fruiting body of a fungus containing ascospores.

Petal—one unit of the inner floral envelope or corolla.

Petiole—leaf stalk.

Pinna (pl. pinnae)—division of a pinnate leaf, such as a fern leaf.

Pistil—female plant reproductive structure consisting of an ovary, style, and stigma.

Plant community—specific plant with its associated organisms in a selected habitat.

Pleistocene epoch—a time period ranging from 1.6 million to 10,000 years before the present when cool glacial periods alternated with warm interglacial intervals.

Pod—a dehiscent (with valves) dry fruit containing seeds.

Pollen—male reproductive element of a flower, corresponding to sperm in animals.

Pollinator—an agent, usually an insect, that transfers pollen from one flower to another.

Pollinaria—little pollen sacs, especially of an orchid, that become attached to insect pollinators.

Pond—a small body of still water.

Posterior—the tail or tip portion.

Predator—an insect that consumes several prey items during its development (in contrast to a parasite that consumes a single host during its development.

Prehensile—capable of grasping.

Pronotum—a hardened structure covering the prothorax.

Prostrate—lying flat or growing along the surface of the ground.

Pterostigma—an opaque thickened spot on the upper end of the wing of some insects.

Pupa (pl. pupae)—a quiescent developmental stage between the larva and adult insect.

Puparium (pl. puparia)—pupal case made by hardening of the larval skin as in many flies.

Pycnidia—asexual fruiting body of a fungus containing spores.

Regional habitat—range and habitat of an organism living in the Pacific Northwest dune forests and wetlands.

Reticulate—network-like.

Rhizome—underground stem that produces aerial stems as it grows.

Rivulet—a very small source of flowing water.

Runner—a long plant stem that grows along the ground. Also called a stolon.

Samara—a winged fruit containing a single seed.

Sand dune—a wide expanse of sand in the form of a mound that can extend for long distances.

Saprophyte—an organism that obtains nourishment from dead organic material.

Scabies—a skin condition marked by prickling small eruptions caused by itch mites.

Scavenger—animal that feeds on dead animal and plant remains.

Seed—a ripened ovule usually borne in a fruit that many be fleshy (apple) or dry (legume pod).

Seep—area where water slowly seeps out of the ground and collects for a short distance.

Sepal—one of the elements of the calyx on a flower.

Seta (pl. setae)—sharp, stiff hair on the surface of an object.

Shore—land along the edge of a large body of water.

Shrub—a woody perennial that remains low and produces stems from the base.

Solitary—single or alone.

Spadix—a cluster of tiny flowers positioned on an erect axis, often protected by a spathe.

Spathe—a large bract that shields flowers from physical damage.

Species—a group of interbreeding individuals with similar characteristics.

Spike—simple, unbranched cluster of sessile flowers.

Spine—sharp, pointed, woody structure arising from the stem.

Spiracle—breathing pore on the sides of the insect abdomen.

Spittle—the frothy deposit produced by spittlebug nymphs.

Springtail—small, wingless hexapod with a spring-like organ (furcula) under its abdomen that is used for leaping.

Spore—a simple reproductive body found in fungi and ferns.

Stamen—male plant reproductive structure consisting of a filament and attached anther.

Stem—the main leaf-bearing and flower-bearing axis of a plant.

Sterigmata—small projection attaching a basidiospore to a basidium.

Stigma—that part of the pistil that receives the pollen.

Stolon—a long plant stem that grows along the ground. Also called a runner.

Stylet—a pointed structure one organism inserts into another for obtaining fluids.

Swamp—a permanent wet area containing aquatic vegetation.

Symbiont—an organism that forms an association with a different organism.

Symbiosis—an association of two organisms that can benefit both or harm one.

Temporary pond—still water body formed during the rainy season with the water gradually evaporating by midsummer.

Thorax—structural part of an insect between the head and abdomen, where the legs and wings are attached.

Trifoliate—with leaves divided in three parts.

Vector—usually a biting insect or arachnid, such as a mosquito or tick that transmits pathogens from one organism to another.

Venation—arrangement of veins in a leaf.

Wetlands—see "dune wetlands."

Whorl—three or more leaves or flowers emerging from the same node on the stem.

Wing covers—elytra or hard forewings on beetles.

SELECT REFERENCES

Arora, D. 1986. Mushrooms demystified. A comprehensive guide to the fleshy fungi. Ten Speed Press, Berkeley, California, U.S.A. 959 pp.

Essig, E.O. 1958. Insects and mites of Western North America. MacMillan Co., New York, U.S.A. 1050 pp.

Haggard, P. & J. Haggard. 2006. Insects of the Pacific Northwest. Timber Press, Inc., Portland, Oregon, U.S.A. 295 pp.

Hatch, M. 1957–1971. The beetles of the Pacific Northwest, 6 vols. University of Washington Press, Seattle, Washington, U.S.A.

Hickman, J.C., ed. 1996. The Jepson manual: Higher plants of California. University of California Press, Berkeley, California, U.S.A.1400 pp.

Hitchcock, C.L. & A. Cronquist. 1973. Flora of the Pacific Northwest. University of Washington Press, Seattle, Washington, U.S.A. 730 pp.

Meyers, S.C., T. Jaster, K.E. Mitchell, & L.K. Hardison, eds. 2015. Flora of Oregon. Vol. 1: Pteridophytes, gymnosperms, and monocots. Botanical Research Institute of Texas Press, Fort Worth, Texas, U.S.A. 592 pp.

Miller, J.C. & P.C. Hammond. 2003. Lepidoptera of the Pacific Northwest: Caterpillars and adults. US department of Agriculture. Forest Health Technology Enterprise Team, Morgantown, West Virginia, U.S.A. 324 pp.

Moore, M. 1993. Medicinal plants of the Pacific West. Museum of New Mexico Press, Santa Fe, New Mexico, U.S.A. 359 pp.

Muenscher, W.C. 1951. Poisonous plants of the United States. The Macmillan Company, New York, U.S.A. 277 pp.

Poinar, Jr., G.O. 2016. A naturalist's guide to the hidden world of Pacific Northwest dunes. Oregon State University Press, Corvallis, Oregon, U.S.A. 286 pp.

Pojar, J. & A. MacKinnon. 1994. Plants of the Pacific Northwest coast. Lone Pine Publishing, Redmond, Washington, U.S.A. 527 pp.

Powell, J.A. & C.L. Hogue. 1979. California insects. University of California Press, Berkeley, California, U.S.A. 388 pp.

Powell, J.A. & P.A. Opler. 2009. Moths of Western North America, University of California Press, Berkeley, California, U.S.A. 369 pp.

Turner, M. & P. Gustafson. 2006. Wildflowers of the Pacific Northwest. Timber Press, Inc., Portland, Oregon, U.S.A. 511 pp.

INDEX

Note: In items with multiple listings, page numbers in **bold** refer to figures and accompanying descriptions.

D